Copernicus Books
Sparking Curiosity and Explaining the World

Drawing inspiration from their Renaissance namesake, Copernicus books revolve around scientific curiosity and discovery. Authored by experts from around the world, our books strive to break down barriers and make scientific knowledge more accessible to the public, tackling modern concepts and technologies in a nontechnical and engaging way. Copernicus books are always written with the lay reader in mind, offering introductory forays into different fields to show how the world of science is transforming our daily lives. From astronomy to medicine, business to biology, you will find herein an enriching collection of literature that answers your questions and inspires you to ask even more.

Sam Goldstein · Robert B. Brooks

The Power of Resilience for Autistic Adults

Thriving in a Neurotypical World

Sam Goldstein
School of Medicine
University of Utah
Salt Lake City, UT, USA

Robert B. Brooks
Former Faculty, Department of Psychiatry
Harvard Medical School
Needham, MA, USA

ISSN 2731-8982 ISSN 2731-8990 (electronic)
Copernicus Books
ISBN 978-3-032-09959-4 ISBN 978-3-032-09960-0 (eBook)
https://doi.org/10.1007/978-3-032-09960-0

© The Editor(s) (if applicable) and The Author(s), under exclusive license to Springer Nature Switzerland AG 2026

This work is subject to copyright. All rights are solely and exclusively licensed by the Publisher, whether the whole or part of the material is concerned, specifically the rights of translation, reprinting, reuse of illustrations, recitation, broadcasting, reproduction on microfilms or in any other physical way, and transmission or information storage and retrieval, electronic adaptation, computer software, or by similar or dissimilar methodology now known or hereafter developed.
The use of general descriptive names, registered names, trademarks, service marks, etc. in this publication does not imply, even in the absence of a specific statement, that such names are exempt from the relevant protective laws and regulations and therefore free for general use.
The publisher, the authors and the editors are safe to assume that the advice and information in this book are believed to be true and accurate at the date of publication. Neither the publisher nor the authors or the editors give a warranty, expressed or implied, with respect to the material contained herein or for any errors or omissions that may have been made. The publisher remains neutral with regard to jurisdictional claims in published maps and institutional affiliations.

This Springer imprint is published by the registered company Springer Nature Switzerland AG
The registered company address is: Gewerbestrasse 11, 6330 Cham, Switzerland

If disposing of this product, please recycle the paper.

Autism can't define me. I define autism.
Kerry Magro

If you've met one individual with autism, you've met one individual with autism.
Stephen Shore

I am different, not less.
Temple Grandin

Autism is not a disease. Don't try to cure us. Try to understand us.
Brian R. King

This book is for adults with autism spectrum disorder who are navigating a world not always built for them. May you find strength in your journey, confidence in your abilities, and the unwavering belief that you are enough, exactly as you are. Your resilience inspires us all.

And to my beloved wife, Sherrie, our children, and our grandchildren. Your love, support, and encouragement make every challenge worth facing and every triumph even more meaningful.

Sam Goldstein

To my supervisors and mentors for offering their wisdom and encouragement during my career, to my close friend and colleague Sam Goldstein for our truly enriching collaboration, and to the many children, teens, adults, and families I have seen in my clinical practice for teaching me the importance of focusing on their strengths and resilience.

And to my immediate family Marilyn, Rich, Doug, Gigi, Suzanne, Maya, Teddy, Sophie, and Lyla for providing countless moments of love and support that have impacted both my personal and professional lives.

Robert B. Brooks

We also wish to thank Judy Jones, our longtime Editor at Springer, for her unwavering dedication, sharp editorial judgment, and steady leadership that has elevated our work year after year.

Sam Goldstein

Robert B. Brooks

Preface

Life is a mosaic of challenges and triumphs, a tapestry shaped by our choices and the resilience we cultivate. In today's world, this tapestry has become increasingly intricate, woven together with the rapid pace of technological advancement, shifting social dynamics, and the ever-expanding expectations of a hyper-connected society. For adults with autism spectrum disorder (ASD), the journey often involves navigating a world designed around neurotypical norms, which can amplify the inherent complexities of relationships, careers, and self-identity.

The pressure to conform to societal expectations, the demand for constant communication, and the blurred boundaries between personal and professional life can create unique obstacles. Thriving in such an environment requires an extraordinary blend of adaptability, self-awareness, and perseverance—a concept encapsulated by the term "resilience." This book explores that journey, highlighting how resilience can empower individuals with ASD to carve out spaces of belonging, redefine success on their own terms, and harness their strengths in a world that often overlooks neurodiversity.

Before proceeding, please note that throughout this book we shift between identity-first language (i.e., autistic adults) and person-first language (i.e., adults with autism). On occasion, we also use "on the spectrum." We recognize that there is a growing preference for identity-first language, but many still adhere to person-first language, while others remain more comfortable with the words "on the spectrum." Our use of these different languages represents an acknowledgment of these various perspectives. However, whatever the language used, we advocate not losing sight of the fact that in addition to the thoughts and behaviors that qualify for a diagnosis, all individuals possess different interests and strengths that can be nurtured to help them thrive in a neurotypical world.

The Complex Impact of ASD

ASD is a lifelong neurodevelopmental condition that influences social communication and behavior. While much research has been dedicated to understanding ASD in children, the health status of adults with ASD has received significantly less attention. The complex challenges this population faces—including physical and mental health conditions and barriers to accessing healthcare—have a broader impact on their quality of life.

Many autistic adults face challenges in finding and keeping jobs due to difficulties with social communication, sensory sensitivities, and rigid thinking patterns. Job interviews can be especially difficult since they often require strong interpersonal skills and the ability to understand social cues. Once in a job, workplace discrimination andand a lack of accommodations present additional obstacles. Employers may not be aware of the specific needs related to autism, a situation that results in insufficient support systems. Many autistic employees report facing discrimination, bullying, or being passed over for promotions, all of which make it even more challenging to succeed in professional environments.

And even when individuals secure jobs, job retention remains a significant issue. Workplace environments that lack structure or present unpredictable changes can be overwhelming. High levels of anxiety and burnout due to sensory overload or social demands contribute to frequent job turnover, making long-term stability difficult for many autistic adults. Maintaining employment becomes a persistent challenge without appropriate accommodations and understanding from employers.

Family life presents its own set of challenges for autistic adults. Relationship difficulties often arise from issues in understanding and expressing emotions, which can complicate romantic relationships and parenting, and misreading social cues can lead to misunderstandings and conflicts with family members, further straining those relationships. Moreover, families of autistic adults often face a considerable emotional and financial burden due to caregiving responsibilities. Aging parents in particular are concerned about long-term care and the independence of their autistic children. This concern increases stress in their daily lives.

Social isolation also remains a significant concern. Many autistic adults struggle to form and sustain friendships, resulting in feelings of loneliness and exclusion. Family members may also experience isolation due to insufficient understanding and support from their community. The cumulative effects of workplace challenges, family pressures, and social isolation can make daily life especially difficult for autistic adults. This reality underscores the need for better awareness, accommodations, and support systems.

Autistic adults are more likely to experience chronic illnesses and immune-related disorders that can significantly affect their daily lives. Epilepsy is one of the most prevalent conditions and is often linked to cognitive and behavioral difficulties. Gastrointestinal issues (including constipation and irritable bowel syndrome or IBS) are also more common among individuals with ASD, contributing to discomfort and poor health. Sleep disturbances—particularly circadian rhythm sleep-wake disorders—are frequently reported, with many experiencing delayed sleep phase disorder. Anxiety, depression, and certain medications further exacerbate these sleep issues. Additionally, obesity, hypertension, and type 2 diabetes appear more frequently in autistic adults, often due to a combination of dietary habits, physical inactivity, and medication side effects. Immune system disorders such as asthma and allergies and autoimmune diseases like type 1 diabetes and psoriasis are also more prevalent in this population. Some individuals with ASD have an increased sensitivity to pain, while others may have a diminished pain response. This variation in sensory processing can make it difficult for them to communicate discomfort, leading to undiagnosed or untreated health issues. Understanding these differences is crucial for providing appropriate medical care tailored to the needs of individuals with ASD.

Mental health disorders are also a significant concern—autistic adults experience anxiety and depression at much higher rates than the general population. Social isolation, employment difficulties, and executive functioning challenges contribute to these conditions.

Many individuals with ASD also face challenges associated with obsessive-compulsive disorder (OCD), which appears as repetitive behaviors and rigid thought patterns. Some studies indicate a higher prevalence of schizophrenia and other psychotic disorders among autistic adults, whereas social anxiety remains a significant obstacle to building relationships and engaging in social interactions.

The presence of multiple co-occurring conditions—both physical and mental—has profound consequences for autistic adults. Many require substantial medical and psychological support but often face difficulties in accessing appropriate care. Research indicates that autistic adults have a lower health-related quality of life compared to their neurotypical peers and usually need assistance with daily activities, particularly those who also have intellectual disabilities.

But despite their high healthcare needs, autistic adults face significant barriers when seeking medical care. Communication difficulties can make it challenging to accurately express symptoms, which may lead to misdiagnoses or inadequate treatment. Sensory sensitivities can make medical examinations stressful and discourage individuals from pursuing care. Furthermore, many healthcare professionals lack specialized training in ASD, and this knowledge gap can hinder optimal medical management. Systemic and financial issues create additional obstacles, including limited insurance coverage and a shortage of ASD-specialized services.

This data underscores the urgent need for educational, vocational, medical, and mental healthcare approaches specifically tailored to the needs of autistic adults.

Resilience as a Beacon of Hope

The inspiration for *The Power of Resilience for Autistic Adults* comes from decades of clinical practice, research, and personal experiences with individuals who exemplify the courage to thrive against the odds. Throughout the years, we have worked with countless children and adults on the spectrum—individuals striving to build fulfilling lives despite needing to surmount barriers that are often invisible to others. Whether it's a young professional managing sensory sensitivities in a bustling workplace, an autistic parent balancing the demands of family life, or an entrepreneur harnessing their intense focus into innovation, each person's story underscores a universal truth: resilience is not innate but must be nurtured.

For autistic adults, resilience is particularly crucial in the face of well-documented challenges and risks. Research has shown that autistic adults experience significantly lower rates of employment compared to their neurotypical peers, with estimates suggesting that only 15% to 20% hold full-time jobs. Workplace discrimination and a lack of accommodations exacerbate these struggles, leading to high turnover rates and chronic underemployment. Sensory sensitivities, difficulties with unstructured work environments, and social communication barriers further complicate professional success. Additionally, autistic individuals are at a higher risk for mental health conditions such as anxiety and depression, often stemming from workplace stress, social isolation, and difficulties in navigating daily life.

Family life also presents obstacles, as autistic adults frequently encounter difficulties with initiating and sustaining romantic relationships, parenting, and making social connections. Studies indicate that marriages and partnerships involving an autistic individual often require additional communication support as emotional expression and social cues can be misinterpreted. Parents on the spectrum may struggle with the unpredictabilities of family life—they must balance their personal needs with the demands of caregiving. At the same time, families of autistic adults face substantial emotional and financial pressures, particularly when it comes to parents worrying about their children's long-term independence and support systems.

Social isolation is another major concern—studies highlight that many autistic adults report loneliness and a lack of meaningful friendships. Without strong support networks, the risk of mental health struggles increases, further reinforcing the cycle of isolation. The combined effects of workplace barriers, family pressures, and social disconnection underscore the necessity of building resilience.

In the face of all of these challenges, we wrote this book to serve as a guide and companion for autistic adults, their families, and the professionals who support them. The strategies and insights we provide here offer practical tools and affirm each reader's value and potential. Resilience is not merely a skill, it's a mindset that empowers individuals to reshape their narratives, embrace their strengths, and live authentically. By fostering resilience, autistic adults can navigate a world that may not always meet their needs while creating spaces where they can thrive, contribute meaningfully, and define success on their own terms.

Bridging the Gap: Why Now?

In recent years, awareness of autism has grown significantly, yet much of the dialogue still focuses on children. As these children transition into adulthood, a critical gap in support and understanding becomes apparent—autistic adults face unique challenges, from building relationships to securing meaningful employment, yet resources tailored to their needs remain scarce. Meanwhile, societal discussions about neurodiversity and mental health have generated new opportunities for inclusion and acceptance, making this an ideal moment to emphasize the resilience of autistic adults. However, these discussions have also sparked controversy and debates about how to best support individuals on the spectrum.

Some advocate for better accommodations and systemic changes that would enable neurodivergent individuals to thrive on their own terms, while others emphasize the importance of therapies and interventions aimed at helping those with ASD integrate into a predominantly neurotypical world. The conflict between these perspectives has led to discussions about autonomy, identity, and the ethics of intervention. Many autistic adults reject the notion that they need to be "fixed"; instead, they promote environments that celebrate neurodiversity rather than impose conformity.

Amid this landscape, the need for this book is urgent. Many autistic adults were diagnosed later in life or missed out on early interventions tailored to their needs. They may bear the weight of years spent feeling misunderstood and navigating a world that often seems unwelcoming. This book aims to meet these individuals where they are by providing them with tools to understand and overcome past obstacles while envisioning a future filled with possibilities. It acknowledges the complexities of the ongoing debates while focusing on one constant that unites all perspectives: the power of resilience in helping individuals with ASD build meaningful, fulfilling lives.

A Strengths-Based Approach

What sets this book apart is its unwavering focus on strengths. Too often, discussions about autism center around correcting deficits or challenges, neglecting the unique abilities and potential of individuals on the spectrum. In contrast, we view ASD as a variation of the human experience, where strengths such as creativity, focus, and perseverance often shine through. When nurtured, these qualities can become powerful assets in both personal and professional lives, enabling individuals with ASD to forge their own paths instead of merely adapting to the expectations of a neurotypical world.

Each chapter is designed to help readers build on their strengths while addressing specific

aspects of resilience, ranging from managing anxiety and stress to cultivating empathy and enhancing communication skills. We recognize that challenges exist, but we approach them believing that self-awareness and self-advocacy are essential for overcoming barriers. Our goal is not to ask readers to conform to neurotypical expectations but to empower them to flourish in a manner that aligns with their authenticity.

Resilience is the bridge between potential and fulfillment. It is not about masking differences but rather embracing them as integral aspects of identity. By shifting the conversation from limitations to empowerment, we aim to provide a roadmap for autistic adults to harness their strengths, navigate challenges with confidence, and build lives rich in purpose and self-acceptance.

For Readers and Allies Alike

This book is designed for autistic adults who are ready to embrace the challenge of building resilience, whether they have long acknowledged their diagnosis or are just beginning to understand how it affects their experiences. It also serves as a resource for their families, friends, and professionals who aim to better understand and support them. Many loved ones and caregivers want to provide meaningful assistance but may find it challenging to grasp the unique ways that ASD impacts daily life. This book equips them with insights and tools to foster deeper connections, respect individual needs, and create nurturing environments where neurodivergent individuals can thrive.

By integrating psychological principles, real-life examples, and practical strategies, we strive to foster greater empathy, connection, and empowerment for all those affected by ASD. Resilience encompasses not only personal strength but also the presence of a supportive and understanding community. Through education and shared experiences, we work to close the gap between neurodivergent and neurotypical perspectives; we aim to promote acceptance and advocacy and to redefine success to celebrate the diversity of the human mind.

A History of Collaboration

At this point, we should introduce ourselves and share the journey that brought us together to write this book. Our combined expertise in psychology and developmental support—especially in assisting autistic children and adults and other developmental challenges—has been shaped by a shared vision: to create meaningful, lasting change in the lives of individuals and their families. For over 30 years, we have worked side by side, continually refining our approach and expanding the boundaries of what traditional psychological care can offer.

Our collaboration began out of a mutual desire to move beyond the standard models of care that existed when we first met. Those models often focused on diagnosing deficits and addressing weaknesses, but we saw an opportunity to shift the focus to building strengths, fostering emotional resilience, and helping individuals with ASD lead fulfilling, empowered lives. By blending our expertise in clinical psychology, education, and therapeutic practices, we developed innovative frameworks that prioritize self-awareness, emotional regulation, and positive social connections.

As professionals who have spent decades working with individuals across the autism spectrum, we've seen firsthand how vital resilience is—not just for those on the spectrum but also for their families. Through our research, workshops, and clinical practice, we've created a body of work aimed at fostering resilience in autistic children and adults. This book is a culmination of that work. It is designed to provide practical strategies for individuals and families and equip them with the tools to thrive, not just survive. We want to share our insights, experiences, and the powerful strategies we've developed so that others may begin their own journeys toward resilience.

Our partnership began in 1992 at a psychology conference, where a conversation about the state of psychological care led to a realization: the deficit-based, "medical model" approach that was so prevalent in our training and practice needed to be reimagined. Instead of focusing solely on diagnosing and "fixing" what was wrong with individuals, we recognized the need to identify and nurture strengths—what we called "islands of competence." We believed this strengths-based approach could help individuals, particularly children with developmental and emotional challenges, develop resilience and a sense of agency.

From that pivotal moment, we embarked on a collaborative effort that has spanned decades. Our first significant book together, *Raising Resilient Children*, introduced the concept of a "resilient mindset." That has since become a cornerstone of our work. We defined a resilient mindset as a set of beliefs and attitudes that allow individuals to view themselves and the world in a way that increases their growth, adaptability, and perseverance. This mindset, we argued, was not something that people either had or didn't have; instead, it was a skill that could be cultivated over time through intentional practice and suitable support systems.

This concept of resilience has been fundamental in the context of our work with autistic children. Many of these children face significant challenges in communication, social interaction, and emotional regulation. Traditional approaches often focused on their deficits—what they couldn't do—rather than on their strengths and the unique abilities they possessed. We sought to change that narrative; we emphasized the importance of recognizing and building on these children's inherent strengths, whether in specific interests, intellectual skills, or creative talents.

Our belief in the power of resilience also led us to explore how it intersects with other essential aspects of personal development, such as self-discipline. We observed that autistic children often faced difficulties in regulating their emotions and behaviors, a challenge that could be daunting for both the children and their families. As we continued our clinical work, we recognized that self-discipline—the ability to manage one's actions, impulses, and emotional responses—was vital for long-term success and emotional well-being. This realization prompted us to co-author *Raising a Self-Disciplined Child*, where we examined the links between resilience and self-discipline and how nurturing both qualities in children (including those with ASD) could equip them with the tools to navigate life's challenges more effectively.

Throughout our collaboration, we have been committed to making our research and insights accessible to parents, educators, and professionals who work with children facing developmental challenges. Our work has expanded beyond theory and now encompasses practical strategies that can be applied in everyday life. As we noted earlier, one of the most critical aspects of this approach has been helping parents and educators understand the importance of focusing on a child's strengths rather than solely addressing their deficits.

Our work has also underscored the significance of creating environments where autistic children can flourish. Resilience is not merely an individual trait! On the contrary, it can be nurtured through supportive relationships and settings. Consistent support and understanding from parents, teachers, and caregivers are crucial for helping autistic children develop resilience. We have emphasized the necessity for caregivers to establish nurturing environments that strengthen children's abilities while equipping them with coping strategies for their challenges.

A crucial part of our message has been the role of "charismatic adults" in a child's life, a term first introduced by the late psychologist Julius Segal. These are adults from whom children "gather strength"—adults who provide encouragement, guidance, and belief in the child's potential. We have often asserted that resilience is rooted in relationships, and for autistic children, having a caring, understanding adult in their lives can make a world of difference. This adult could be a parent, teacher, coach, or therapist who helps the child face challenges while reinforcing their strengths. As we noted in our book *Raising Resilient Children with Autism Spectrum Disorders*, we have seen time and time again that when children with ASD are supported by a charismatic adult, they can develop the resilience they need to overcome significant obstacles and find hope and fulfillment in their lives.

As part of our work, we've also developed tools and strategies specifically designed for autistic children to help them manage their emotional and behavioral challenges. These include techniques for emotional regulation, such as mindfulness and breathing exercises that help children recognize and cope with their emotional triggers. We've also emphasized the importance of setting realistic goals and breaking down larger tasks into smaller, more manageable steps to help children on the spectrum build self-confidence and experience a sense of mastery over their environment.

In our most recent projects, we have continued to focus on resilience in children with ASD and other challenging conditions like Disruptive Mood Dysregulation Disorder (DMDD), particularly with respect to how they can be empowered to take control of their own lives. We've witnessed the transformative power of resilience in helping autistic children not only cope with their challenges but also develop problem-solving skills that reinforce a sense of autonomy and independence. One of our core messages is that resilience is not about eliminating difficulties, it's about equipping children with the tools they need to face those obstacles with confidence and adaptability.

Over the years, our collaboration has resulted in the co-authorship of more than 14 books, including *The Power of Resilience: Achieving Balance, Confidence, and Personal Strength in Your Life* and *Nurturing Resilience in Our Children*, along with numerous articles, book chapters, and a radio series. Our parenting programs have been implemented worldwide; they offer parents practical strategies for raising resilient children, especially when those children are facing developmental and emotional challenges. These programs highlight the significance of cultivating a resilient mindset, promoting self-discipline, and creating supportive environments in which autistic children can flourish.

Our collaboration is grounded in the belief that resilience is not a luxury but a necessity for personal growth and well-being. We have seen firsthand how resilience transforms lives—not by eliminating challenges but by providing individuals with the tools to navigate them. This book draws on our combined experience to offer readers scientifically based and deeply human insights, merging research with practical applications.

As we reflect on our work together, we are deeply grateful for having the opportunity to make a meaningful impact on the lives of so many children, adults, and families. This book is more than just a collection of strategies, it's an invitation to reimagine what is possible. It challenges outdated beliefs that resilience is solely about enduring hardships, instead presenting resilience as an active, evolving process of self-discovery and empowerment. For every autistic adult who has felt the weight of societal expectations, every family member seeking to provide meaningful support, and every professional striving to make a difference, this book is our offering. It is a testament to the power of resilience. It is a call to embrace neurodiversity, recognize the value of diverse ways of thinking, and promote an inclusive world where individuals with ASD can thrive on their own terms.

The Path Forward

Resilience is a journey rather than a destination. It's cultivated through small, intentional steps and nurtured by a belief in our ability to grow. For autistic adults, resilience is crucial for unlocking a life of purpose, connection, and fulfillment. It serves as a lens through which challenges become not merely obstacles but opportunities for transformation.

In the chapters ahead, you'll find stories of adaptation and courage, tools for dealing with adversity, and a roadmap for building a resilient mindset. Together, these elements provide a guide to thriving in an ever-changing world that's full of potential. At the end of each chapter, we present five key takeaways followed by practical, self-guided activities designed to help autistic adults engage directly with the concepts we discuss. These activities offer actionable steps for identifying challenges, leveraging strengths, and nurturing personal growth.

Each activity is structured to encourage reflecting, skill-building, and practicing and is tailored to address the unique experiences of individuals with ASD. Whether these activities involve journaling to uncover harmful scripts, role-playing to build confidence in social scenarios, or creating sensory—friendly strategies that can be used on a daily basis, these exercises are designed to meet autistic adults where they're at and guide them toward meaningful progress.

Autistic adults can approach these activities at their own pace, selecting those that resonate most with their current needs and revisiting them as new challenges arise. The exercises are designed to complement the stories and tools provided in each chapter—they turn abstract ideas into tangible actions. Engaging in these activities will provide individuals with insights and practical skills that will help them feel resilient and empowered as they navigate their journeys.

As they embark on this journey, autistic adults need to remember that resilience is *not* about being unshaken—it's about learning to bend without breaking, adapting without losing themselves, and rising stronger with each challenge. Resilience is *not* merely about surviving the demands of a world that often feels unaccommodating—it's about thriving, embracing inner strengths, and crafting a life of fulfillment and purpose. It means recognizing that setbacks are not failures but stepping stones to deeper self-understanding and greater future successes.

Thriving with ASD involves more than just managing challenges! It encompasses embracing unique abilities, finding joy in passions, and forming meaningful connections on one's own terms. This approach transforms the narrative from limitation to empowerment and from struggle to growth. Through resilience, autistic adults can create not only a survivable life, but an extraordinary one that celebrates their authenticity, strengths, and limitless potential.

Welcome to *The Power of Resilience for Autistic Adults*. Let's begin this journey together.

Salt Lake City, USA Sam Goldstein
Needham, USA Robert B. Brooks

A Different Path, A Brighter Light

Through senses sharp and mind so bright,
Autism shapes my day and night.
The world can feel too loud, too tight,
A maze of rules, a shifting tide.

Where spoken words and meanings hide,
Each touch, each sound, a storm inside.
Yet beauty blooms where truths reside,
In patterns deep and logic pure.

A different path, a heart secure,
Challenges arise, but so do I.
With every question, every try,
The world may bend but won't define.

The spark within—so fierce, so fine,
With love and strength, I carve my way.
A voice that shines, a mind that stays,

For being different isn't wrong—
It's where I've found that I belong.

Sam Goldstein

Introduction for Adults with ASD

If you're an adult with ASD, you might wonder whether the strategies and insights offered here truly apply to you. After all, advice about motivation, discipline, or resilience often assumes a neurotypical perspective that may not fully reflect your experience. But know that this book was written with a deep understanding of those differences and—more importantly—with profound respect for them. You are not expected to "fit in" with the world as it is; rather, you are encouraged to build a life that honors who you are on your own terms.

The chapters in this book introduce many challenges you may know well: difficulties in reading social cues, emotional miscommunications, sensory sensitivities, struggles with organization and planning, and the exhausting toll of camouflaging your true self to meet others' expectations. These aren't just occasional hurdles—they are often daily realities that can affect everything from your career to your relationships and self-esteem. That's why some ideas in this book may feel frustrating at first. Suggestions like "stay motivated," "work in groups," or "read nonverbal cues" may sound like things you've heard countless times—without being accompanied by the proper context or support you need to actually succeed. However, this book does not assume that "trying harder" is the answer, especially since we believe that in many situations you *have* tried very hard but haven't seen positive results. Instead, this book aims to help you *work smarter* by adapting strategies that align with your strengths and challenges.

For example, when we write about motivation, we recognize that what some might describe as slow really *is* fast enough for someone with ASD. You may need more time to recover from sensory overload or to plan your next step—and that's okay. Building momentum doesn't mean rushing. It means finding your pace and sticking with it, no matter how gradual that pace may be. A small, consistent routine like reviewing your schedule each morning may have more long-term value than overwhelming yourself with daily to-do lists.

Consider resilience. For some, it means bouncing back from conflict or adversity, while for you, it may involve navigating a work meeting without shutting down in response to the noise or remembering to self-advocate instead of masking discomfort. Resilience isn't about pretending that the world doesn't overwhelm you—it's about developing the tools to *live well regardless*.

It's also important to recognize that strategies such as "rethinking negative scripts" may be more challenging to adopt if you've experienced years of rejection or misunderstanding. You may hold deeply rooted beliefs about being "too much" or "not enough." Changing those beliefs requires time—and support. This book won't rush you through that journey. Instead, we'll provide practical activities and real-life examples to help you rewrite those narratives, one compassionate step at a time.

Above all, this book is about *you*. Not a version of you molded to meet the world's expectations, but the real, authentic person you already are—just with better tools, deeper self-understanding, and more powerful strategies for facing life on the spectrum. Whether you're just beginning to explore your diagnosis or you've lived with it for decades, this book invites you to grow, not by becoming someone else, but by becoming more fully *yourself*.

You deserve that journey. Let's begin it—together.

Contents

1	**The Challenges of ASD: Resilient Mindsets, Negative Scripts, and Personal Control**	**1**
	The Real-World Impact of ASD	3
	Theory of Mind	3
	Joint Attention of Emotion	4
	Planning and Attending to Relevant Details	4
	Understanding the Communicative Content of Gaze	5
	Joint Attention of Behavior	5
	Struggles with Social Information	6
	Understanding Resilience and Autism	6
	The Importance of a Resilient Mindset	9
	Harmful Scripts and Their Impact on Resilience	11
	Coping, Personal Control, and Empowerment	12
	The Path Forward: Building Resilience	14
	Five Key Takeaways	15
	Resilience is Essential for Adults with ASD	15
	Harmful Scripts Can Hinder Growth	15
	Personal Control Leads to Empowerment	15
	Social and Emotional Challenges Require Adaptive Strategies	15
	Support Networks and Self-Advocacy Are Crucial	16
	Self-Guided Activities to Overcome the Challenges of ASD	16
	Activity 1: Identifying Personal Challenges	16
	Activity 2: Recognizing Harmful Scripts	16

	Activity 3: Building Resilience	17
	Activity 4: Enhancing Personal Control	17
	Activity 5: Building Support Networks	17
	Activity 6: Reflecting on Progress	17
2	**Rewriting Negative Scripts for Adults with ASD**	**19**
	Why Individuals with ASD Are Particularly Vulnerable to Negative Scripts.	20
	Cognitive Processing Differences and Rigid Thinking Patterns	20
	Frequent Social Misunderstandings and Rejection	21
	Heightened Sensory Sensitivities	21
	Emotional Regulation Challenges	22
	Societal Expectations and Pressure to Conform	22
	Repetition and the Deep-Rooted Nature of Negative Scripts	23
	The Role of Practice and Repetition	23
	The Impact of Societal Expectations on Negative Scripts	24
	The Role of Self-Compassion in Rewriting Scripts	25
	Breaking the Cycle: Strategies for Long-Term Script Transformation	26
	Navigating Relationships While Reframing Scripts	27
	Overcoming Challenges and Setbacks	28
	Recognizing Harmful Scripts	29
	Reframing Scripts as Empowering Narratives	30
	Addressing Sensory and Emotional Needs	31
	Harnessing Strengths and Interests	32
	Transformative Real-Life Examples	33
	Sophie's Journey to Connection	33
	Michael's Professional Transformation	34
	The Power of Rewriting Negative Scripts	35
	Five Key Takeaways	35
	Self-Guided Activities to Rewrite Negative Scripts	36
3	**Choosing the Path to Become Stress-Hardy Rather than Stressed Out**	**41**
	The Concept of Stress Hardiness	42
	Commitment	42
	Challenge	43
	Personal Control	44
	The Stress of Social and Cognitive Challenges for Adults with ASD	46

The Stress of Understanding the Perspectives of Others	46
The Stress of Emotional Connection	47
The Stress of Planning and Prioritization	48
The Stress of Rapid Social Processing	49
Managing the Stress of ASD-Related Challenges	50
Finding Purpose through Commitment	50
Reframing Challenges as Opportunities	51
Cultivating Personal Control	52
Emotional Regulation and Resilience	53
Integrating Strengths and Addressing Challenges	54
Five Key Takeaways	54
Stress Hardiness as a Mindset	54
The Power of Commitment	54
Reframing Challenges as Growth Opportunities	55
Personal Control Reduces anxiety	55
Emotional Regulation Strengthens Resilience	55
Self-Guided Activities to Choose the Path to Become Stress-Hardy	55
Activity 1: Building Commitment and Finding Purpose and Meaning	55
Activity 2: Reframing Challenges as Opportunities	56
Activity 3: Strengthening Personal Control	56
Activity 4: Cultivating Resilience through Emotional Regulation	56
Activity 5: Enhancing Support Networks	57
Activity 6: Embracing Self-Compassion and Growth	57
Activity 7: Reducing Overwhelm through Simplification	57
Activity 8: Reflecting on Progress	58
4 Viewing Life through the Eyes of Others	**59**

Difficulty with Theory of Mind	60
Impaired Joint Attention of Emotion	60
Executive Function and Empathy Challenges	61
Eye Contact and Emotional Communication Struggles	62
Sensory Processing and Empathy Challenges	62
Problems with Emotional Regulation and Empathy	63
Understanding the Perspectives of Others	63
The Role of Social Stories in Enhancing Empathy	64
Emotional Labeling and Expression	65
Nonverbal Communication and its Impact on Relationships	66
Practicing Empathy in Digital Communications	67
Cultivating Empathy: A Path to Connections	68
Five Key Takeaways	69
Empathy Is a Skill that Can be Developed	69
Challenges with Theory of Mind	69
Nonverbal Cues and Joint Attention Play a Crucial Role in Empathy	69
Sensory Processing and Emotional Regulation Impact Social Engagement	69
Perspective-Taking Is the Key to Meaningful Connections	69
Self-Guided Activities to Cultivate Empathy and Perspective-Taking	70
Activity 1: Engaging in Perspective-Taking Reflection	70
Activity 2: Doing an Empathy-Mapping Exercise	70
Activity 3: Cultivating an Awareness of Sensory Overload	70
Activity 4: Practicing Theory of Mind: "What Are They Thinking?"	71
Activity 5: Having Role-Playing Conversations	71
Activity 6: Writing in an Emotional Regulation Journal	71
5 Effective Communication: Tools for Navigating the Social World	**73**
Why Communication Is So Challenging for People with ASD	74
The Ability to Attribute Mental States to Oneself and Others	75
The Ability to Display Emotional Reactions Appropriate to another Person's Mental State	76
The Ability to Plan and Attend to Relevant Details in the Environment	77
The Ability to Understand the Communicative Content of Gaze	79

	The Ability to Work Cooperatively with Others (Joint Attention of Behavior)	80
	The Ability to Understand, Comprehend, Analyze, Synthesize, Evaluate, and Differentiate Social Information	81
	The Foundations of Effective Communication	83
	Active Listening: Building Bridges of Understanding	83
	Clear Expression: Speaking with Purpose and Precision	84
	Adapting Communication Styles to Social Contexts	85
	The Power of Nonverbal Communication	86
	Navigating Conflicts: Communication as a Tool for Resolution	87
	Building Resilience through Communication	88
	Five Key Takeaways	89
	Explicit Teaching of Social Rules	89
	Literary and Visual Supports	89
	Role-Playing Exercises	90
	Encouraging Self-Advocacy	90
	Mindfulness and Emotional Regulation Techniques	90
	Self-Guided Activities to Improve Communication Skills	91
	Activity 1: Practicing Active Listening	91
	Activity 2: Nonverbal Communication Awareness	91
	Activity 3: Role-Playing Different Social Scenarios	92
	Activity 4: Understanding Social Cues Using Various Media Sources	92
	Activity 5: Practicing Clear and Concise Speech	92
6	**Self-Acceptance and Embracing Individuality**	**95**
	Understanding Self-Acceptance	95
	Challenges to Self-Acceptance and Individuality in Autistic Adults	96
	Difficulty Understanding Mental States	96
	Challenges with Emotional Expression	97
	Difficulties with Planning and Organization	98
	Misinterpretation of Nonverbal Cues	99
	Struggles with Cooperation and Teamwork	100
	Difficulty Navigating Social Norms	101
	Inconsistencies in Social Learning	101
	Impact on Self-Acceptance and Individuality	102
	The Journey toward Self-Acceptance	103
	The Role of Personal Narratives	104
	Reframing One's Narrative	104
	Aligning Actions with Values	104

	The Role of Self-Advocacy	105
	The Connection Between Self-Acceptance and Resilience	106
	The Power of Community and Support	107
	Embracing Authenticity: The Power of Self-Acceptance	108
	Five Key Takeaways	109
	Self-Awareness Builds Self-Acceptance	109
	Reframing Negative Narratives	109
	Setting boundaries Supports Authenticity	109
	Authenticity over Masking	109
	Supportive Communities Foster Growth	110
	Self-Guided Activities to Improve Self-Acceptance	110
	Activity 1: Practicing Daily Self-Affirmations	110
	Activity 2: Reframing Negative Thoughts	110
	Activity 3: Creating a Personal Values List	110
	Activity 4: Practicing Saying no	111
	Activity 5: Finding and Celebrating your Strengths	111
7	**Building Meaningful Connections: The Power of Relationships**	113
	The Foundations of Relationships	114
	Understanding Mental States: A Core Challenge	114
	Emotional Reciprocity: Navigating Joint Attention of Emotion	115
	Attention to Detail and Environmental Awareness	116
	Understanding the Communicative Content of Gaze	116
	Cooperative Behavior and Joint Attention	117
	Processing and Analyzing Social Information	117
	Navigating Challenges	118
	The Role of Social Skills	119
	Nurturing Compassion in Relationships	119
	The Importance of Social Support Networks	120
	Relationships as a Buffer against Stress	120
	Building a Sense of Belonging	121
	Romantic Relationships and ASD	121
	Parent-Child Relationships and ASD	123
	A Lifelong Journey	125
	The Power of Human Connection	125
	Five Key Takeaways	126
	Understanding Individual Social Needs	126
	The Importance of Clear and Direct Communication	126
	Building Emotional Reciprocity and Understanding Social Cues	126

The Role of Routine, Structure, and adaptability	127
The Power of Social Support and a Sense of belonging	127
Self-Guided Activities to Build and Strengthen Relationships	127
Activity 1: Doing a Relationship-Mapping Exercise	127
Activity 2: Practicing Social Scripts	127
Activity 3: Planning Joint Activities	128
Activity 4: Writing in a Communication Reflection Journal	128
Activity 5: Engaging in a Compassion and Understanding Challenge	128

8   Embracing Mistakes: Transforming Challenges into Opportunities for Growth   129

The Unique Challenges of Mistakes for Individuals with ASD	130
The Ability to Attribute Mental States to Oneself and Others	130
The Ability to Display Appropriate Emotional Reactions (Joint Attention of Emotion)	131
The Ability to Plan and Attend to Relevant Details in the Environment	132
The Ability to Understand the Communicative Content of Gaze	133
The Ability to Work Cooperatively with Others (Joint Attention of Behavior)	133
The Ability to Understand, Comprehend, Analyze, Synthesize, Evaluate, and Differentiate Social Information in the Environment	134
Understanding the Fear of Mistakes	134
The Importance of Resilience in Navigating Mistakes	135
Learning from Mistakes: A Path to Growth	136
Strategies for Embracing Mistakes	137
The Role of Self-Compassion in Learning from Mistakes	138
Reframing Mistakes as Valuable Information	139
Building a New Relationship with Mistakes	139
The Lifelong Journey of Growth	140
Five Key Takeaways	141
Mistakes Are Opportunities for Growth	141
The Fear of Mistakes Can Lead to Avoidance	141
Resilience Helps Individuals Navigate Mistakes	141
Self-Compassion Reduces the Negative Impact of Mistakes	141
Reframing Mistakes as Valuable Information Encourages Growth	142
Self-Guided Activities to Manage Mistakes	142

Activity 1: Reframing Negative Thoughts Exercise	142
Activity 2: Role-Playing Social Scenarios	142
Activity 3: Breaking Tasks into Manageable Steps	143
Activity 4: Self-Compassion Meditation or Journaling	143
Activity 5: Creating a "Mistakes and Lessons Learned" Log	143

9 Embracing Success by Focusing on Strengths — 145

Embracing Success Through Developing Cognitive Strengths	145
The Ability to Attribute Mental States to Oneself and Others	146
The Ability to Display Emotional Reactions Appropriate to Another Person's Mental State (Joint Attention of Emotion)	147
The Ability to Plan and Attend to Relevant Details in the Environment	147
The Ability to Understand the Communicative Content of Gaze	148
The Ability to Work Cooperatively with Others (Joint Attention of Behavior)	149
The Ability to Understand, Comprehend, Analyze, Synthesize, Evaluate, and Differentiate Social Information	150
How We Think About Success	151
The Role of Satisfaction	151
Being Proud Without Boasting	152
The Fear of Success	152
Finding Islands of Competence	153
Making Changes Step by Step	154
Celebrating Successes: An ASD Perspective	155
Five Key Takeaways	156
Emphasizing Strengths Builds Confidence	156
Structured Approaches Lead to Success	156
Understanding Social Cues Can Be Learned	156
Confidence in Achievements Shapes Success	157
Celebrating Small Successes Enhances Resilience	157
Self-Guided Activities to Enhance Success	157
Activity 1: Doing a Strengths-Mapping Exercise	157

	Activity 2: Practicing Social Strategies	157
	Activity 3: Developing Tools for Prioritization and Task Planning	158
	Activity 4: Reframing Success by Journaling	158
	Activity 5: Creating a System for Celebrating Rewards	158
10	**Self-Discipline and Control: Strategies for Personal Growth**	**159**
	The Meaning of Self-Discipline	159
	Self-Discipline and Control Challenges for Autistic Individuals	160
	Executive Function Challenges	160
	Emotional Regulation and Control and Social-Cognitive Differences	161
	Contextual Challenges in Applying Self-Discipline	162
	Cognitive Overload and Burnout[Chap:10][ID:Sec6]	163
	Need for Predictability	164
	Why Routines and Habits Matter	165
	Helpful Strategies	166
	Time Management and Prioritizing Tasks	166
	Managing Strong Emotions	167
	Learning from Mistakes[Chap:10][ID:Sec12]	167
	Remaining Motivated	168
	Taking Control of Our Life	169
	Five Key Takeaways	169
	Self-Discipline is a Process—It's not About Perfection	169
	Executive Function Skills Support Self-Regulation	170
	Emotional Control Protects Progress	170
	Predictable Routines Reduce Stress	170
	Reflection Builds Resilience	170
	Self-Guided Activities to Build Self-Discipline and Control	171
	Activity 1: Building a Micro-Routine	171
	Activity 2: Creating an Emotional Regulation Plan	171
	Activity 3: Generalizing One Routine	171
	Activity 4: Tracking Energy and Overload	171
	Activity 5: Making a Motivation Board	172

11 Maintaining Resilience: Creating a Sustainable Resilient Lifestyle — 173
What Does It Mean to Sustain Resilience? — 173
Sustaining Resilience Across Core Social-Cognitive Challenges — 174
  The Ability to Attribute Mental States to Oneself and Others — 175
  The Ability to Display Emotional Reaction Appropriate to Another Person's Mental State — 175
  The Ability to Plan and Attend to Relevant Details in the Environment — 176
  The Ability to Understand the Communicative Content of Gaze — 177
  The Ability to Work Cooperatively with Others (Joint Attention of Behavior) — 177
  The Ability to Understand, Comprehend, Analyze, Synthesize, Evaluate and Differentiate Social Information — 178
Strategies to Sustain Resilience — 179
  The Role of Daily Practices — 179
  Creating Meaningful Routines — 179
  Building a Support Network — 180
  Adapting to Change — 181
  Sustaining Hope and Optimism — 182
A Lifelong Journey — 182
Five Key Takeaways — 183
Resilience is a Daily Commitment, not a One-Time Achievement — 183
Emotional Awareness Must Be Cultivated — 183
Structured Routines Enable Flexibility and Growth — 183
Building Support Networks Strengthens Resilience — 184
Cognitive Tools Can Demystify Social Complexity — 184
Self-Guided Activities to Maintain Resilience — 184
Activity 1: Establishing a Routine of Daily Anchors. — 184
Activity 2: Doing Emotion Mapping After Conversations — 185
Activity 3: Creating a Sensory and Attention-Planning Toolkit — 185

Activity 4: Creating Visual Social Stories or Flowcharts	186
Activity 5: Drilling Yourself on Your Flexibility by Changing Your Routine	186

12 The Path Forward: Thriving as an Adult with Autism — 187
Embracing Authenticity — 187
Staying True to One's Values — 188
Creating Supportive Environments — 189
Redefining Social Connections — 190
Managing Transitions and Change — 191
Practicing Self-Advocacy — 192
Developing One's Strengths — 193
Building Resilience as a Lifelong Process — 194
Journeys with ASD — 195
Five Key Takeaways — 196
Embracing One's Authentic Self is Empowering — 196
Living by One's Values Creates Meaning — 196
Supportive Environments Make a Difference — 196
Social Connections Can Be Redefined — 197
Resilience is Built, not Born — 197
Self-Guided Activities to Thrive as an Adult with Autism — 197
Activity 1: Designing Your Personal Values Map — 197
Activity 2: Doing a Sensory-Friendly Space Audit — 198
Activity 3: Building a Personal Self-Advocacy Script Library — 198

13 Epilogue — 199
Looking Back: A Life Lived on the Spectrum — 199
Childhood: The Silent Struggle — 200
Teen Years: The Masking Begins — 201
College Years: The Balancing Act — 202
Adulthood: A World Built for Others — 203
A Love That Tried, A Love That Ended — 203
Middle Age: The Unraveling — 205
Acceptance: The Road Home — 206
Sixty: A Life Reclaimed — 206

Resources — 209

Index — 213

About the Authors

Sam Goldstein (PhD) is a neuropsychologist specializing in school psychology, child development, and neuropsychology. Licensed in Utah, he is a certified Developmental Disabilities evaluator, board-certified Pediatric Neuropsychologist, and a Fellow of leading neuropsychology and cerebral palsy academies. He has served as Assistant Clinical Instructor in Psychiatry and has directed a private multidisciplinary team addressing neurological, learning, and behavioral issues since 1980. Dr. Goldstein is on staff at the University Neuropsychiatric Institute and has contributed to the Craniofacial Team and Developmental Disabilities Clinic at the University of Utah. A prolific author, Dr. Goldstein has written or edited over 50 publications, including 21 textbooks on behavior, ADHD, resilience, and learning disabilities, and the Clinician Guide to Disruptive Mood Dysregulation Disorder (2024). He has developed a dozen psychological tests. He has also collaborated extensively, producing seminal works on resilience, classroom management, autism, and neurodevelopmental disorders. Visit samgoldstein.com.

Robert B. Brooks (PhD) is a clinical psychologist who has served on the faculty of Harvard Medical School and is the former Director of Psychology at McLean Hospital. He is an expert on resilience, education, special needs, psychotherapy, parenting, and fostering positive environments. Dr. Brooks has authored or co-authored 23 books, 36 book chapters, and 36 peer-reviewed articles. He has received numerous awards, including a Hall of Fame

Award from the Connecticut Association of Children with LD and CHADD, the Mental Health Humanitarian Award from William James College, and the Trailblazer Award from Worldmaker International for his contributions to the field of resilience. Dr. Brooks has also consulted to Sesame Street Parents Magazine and continues to inspire through his lectures, writings, and groundbreaking work in mental health. Visit www.drrobertbrooks.com.

Also By These Authors:

Raising Resilient Children (2001)
Seven Steps to Help Your Child Worry Less (with Kristy Hagar) (2002)
Parenting Resilient Children Parent Training Manual (2002)
Nurturing Resilience in Our Children (2003)
The Power of Resilience (2004)
Angry Children, Worried Parents (with Sharon Weiss) (2004)
Handbook of Resilience in Children (2005)
Seven Steps to Improve Your Child's Social Skills (with Kristy Hagar) (2006)
Understanding and Managing Children's Classroom Behavior – 2nd Edition (2007)
Raising a Self-Disciplined Child (2009)
Raising Resilient Children with Autism Spectrum Disorders (2012)
Handbook of Resilience in Children – 2nd Edition (2012)
Play Therapy Interventions to Enhance Resilience (with David Crenshaw) (2015)
Tenacity in Children (2021)
Handbook of Resilience in Children – 3rd Edition (2023)
Graham and Poppy's Quest for Tenacity (2024)
Finding the Calm Child Within (with Donna Rooney and Molly Anthony) (2025)

1

The Challenges of ASD: Resilient Mindsets, Negative Scripts, and Personal Control

ASD presents a broad range of challenges for adults, especially in a society that often prioritizes social interactions, conformity, and adherence to established norms. For autistic adults, this is not a revelation—this has always been their reality. The struggles, nuances, and continuous need to navigate through a world built for others are experiences they've likely encountered for years. However, for loved ones, family members, and professionals, this viewpoint may provide new insights into the daily lives of adults with ASD.

Autistic adults frequently encounter hurdles related to communications and social interactions, managing sensory sensitivities, and regulating emotions. For adults on the autism spectrum, the challenges of securing and maintaining employment, building meaningful relationships, and navigating complex social environments can often feel overwhelming, even if they aren't new. Yet despite the scale of these challenges, resilience—the capability to adapt, persevere, and thrive in the face of adversity—is not only achievable but also often a testament to the strength and adaptability of adults with ASD. The depth and intensity of the ongoing difficulties these adults face may not be fully appreciated by others. For those who are supporting autistic individuals, acknowledging these challenges can be a vital step toward fostering enhanced understanding, empathy, and support.

People with ASD are, first and foremost, individuals with unique experiences, strengths, and aspirations. Their challenges, though often shaped by ASD, do not define their humanity. Just like anyone else, autistic adults have the ability and need to build resilience. However, they must approach developing a resilient mindset through the lens of their specific social and pragmatic challenges. Building resilience for autistic adults involves navigating a world that can often be overwhelming, in both a social sense and a sensory sense; for them, developing resilience is *not* about changing who they are but finding ways to thrive in their environment. For example, resilience for a neurotypical individual might involve handling social conflicts. In contrast, for someone with ASD, it could include managing sensory overload in a crowded space or decoding the unspoken rules of social interactions.

Resilience can be cultivated through personal strategies like self-awareness, external support systems, and adaptive coping mechanisms. For many adults with ASD, this means leveraging their unique strengths—such as attention to detail, ability to maintain focus, or specialized interests—while developing practical tools to manage social difficulties. With the right strategies in place, autistic adults can achieve personal and family fulfillment, independence, and professional success.

This chapter explores the concept of resilience as it relates to adults with ASD, emphasizing its essential role in managing social, personal, and professional challenges. For autistic individuals and their supporters, recognizing the significance of a resilient mindset that's tailored to their unique experiences is vital for encouraging positive growth and success.

In that spirit, in this chapter, we outline practical strategies and resources that promote resilience, empowering individuals to thrive despite their specific challenges. These strategies include developing self-awareness, mastering emotional regulation techniques, and connecting with supportive networks that understand the unique difficulties faced by individuals with ASD. By applying these methods, autistic adults can learn not only to manage adversity but also to become stronger, more adaptable, and more capable of achieving their goals. We intend to inspire readers by demonstrating that resilience is not an inherent quality but a skill that can be cultivated over time. This truth provides hope and practical guidance for a more empowered future.

The Real-World Impact of ASD

Autistic adults face a variety of challenges that affect many areas of daily life; these often stem from their difficulties in understanding and navigating social environments. Such fundamental issues shape how individuals with ASD interact with the world and manage tasks that many neurotypical individuals might take for granted. Recognizing these core issues provides a foundational perspective on the wide-ranging impacts that ASD has on adult life. Each of these areas is interconnected, influencing everything from social relationships to professional success, and each of these areas offers insight into the complexities of life on the spectrum.

Theory of Mind

One of the most significant challenges for autistic adults is the ability to attribute mental states to themselves and to others, often referred to as "theory of mind." This ability allows individuals to understand that other people have thoughts, feelings, and intentions that may differ from their own. For autistic adults, difficulties in this area can make it hard to interpret social cues or predict how others might react in certain situations. This can lead to misunderstandings in communication and difficulty in forming and maintaining relationships. In professional or personal settings, individuals may struggle to anticipate the needs of others or adjust their behavior to fit social expectations, leading to feelings of isolation or frustration.

James, a 28-year-old with ASD, attended a team meeting at work and was eager to present a new project idea. During his presentation, one of his colleagues, Ann, began to fidget and glance at her watch, which was a not-very-subtle sign that she might be losing interest. However, James didn't notice her nonverbal cues and continued to explain every project detail enthusiastically. After the meeting, James was surprised to learn that Ann and others had found the presentation to be too long and unengaging. Since he struggled with understanding their perspectives, he hadn't realized that adjusting the length of his presentation or asking for questions or comments midway through could have improved the experience for his audience. This led to frustration for James and his colleagues, creating barriers in communication and making it difficult for him to form stronger workplace relationships. His inability to interpret social cues like Ann's body language highlight one of the challenges that many autistic adults face in social and professional interactions.

Joint Attention of Emotion

Another core issue involves displaying emotional reactions appropriate to another person's mental state, which we describe as "joint attention of emotion." This refers to the capacity to share or acknowledge the emotional experiences of others, a crucial aspect of empathy. For adults with ASD, this can be a particularly challenging area, as they may struggle to interpret the emotions of others or to respond in ways that seem socially appropriate. Sometimes they may appear indifferent or disconnected, even when they care deeply. This can create barriers in both personal and professional relationships, as others may misinterpret their emotional responses. The disconnect between internal feelings and outward expressions of emotion can complicate interactions and create further isolation.

John, a 40-year-old man with ASD, often faced challenges in his marriage due to misunderstandings about his emotional responses. One evening, his spouse, Sarah, came home visibly upset after a difficult day at work. While John noticed her distress, he struggled to respond in a way that Sarah expected. Instead of offering comfort, he remained silent, unsure how to approach the situation. Internally, John felt concerned and wanted to help, but Sarah perceived his silence as indifference. Hurt by what she saw as a lack of empathy, Sarah felt more alone in her struggles. This disconnect led to tension in their relationship, as John's difficulty in expressing "joint attention of emotion" unintentionally caused emotional distance.

Planning and Attending to Relevant Details

The ability to plan and attend to relevant details in the environment is another critical issue for autistic adults. Often, these individuals focus intensely on specific details but may struggle to prioritize or organize tasks efficiently. This challenge can affect their ability to manage daily routines, navigate work responsibilities, or engage in long-term planning. As a result, they may feel overwhelmed or misunderstood in environments that require multitasking or flexible thinking, such as in many workplaces. Nevertheless, this attention to detail can also be a strength in certain fields that demand precision and a focus on intricate tasks, like technology or research.

Understanding the Communicative Content of Gaze

Understanding the communicative content of gaze is another area where adults with ASD often face challenges. Eye contact, a common and powerful social signal, is frequently difficult for autistic adults. While some may avoid eye contact altogether, others may struggle to interpret its meaning— does it indicate interest, disinterest, discomfort, or connection? This can create significant barriers to effective communication, as so much human interaction is conveyed through nonverbal cues like eye contact. Difficulty in interpreting gaze coupled with challenges in reading body language can make social interactions more taxing and produce misunderstandings that further isolate adults with ASD.

Joint Attention of Behavior

Working cooperatively with others, often referred to as "joint attention of behavior," presents a significant challenge. Collaborative efforts usually require sharing focus with another person and responding to their actions or directions in real time. For autistic adults, this can be a daunting task, especially given the presence of unspoken social rules or ambiguous cues. In the workplace, this may manifest as difficulty participating in team projects, misunderstandings with colleagues, or challenges in meeting supervisors' expectations. While individuals with ASD may thrive in independent tasks, the subtleties of collaborative work can lead to stress and confusion.

Alan, a 42-year-old with ASD, recently joined his neighborhood's Homeowners Association (HOA) committee to help improve community spaces. Although he was enthusiastic about contributing his ideas, working with other homeowners on the committee presented several challenges. During meetings, Alan struggled to follow the rapid back-and-forth discussions and the subtle nonverbal cues the other members were exchanging. When one committee member suggested a different approach to an issue Alan was passionate about, Alan failed to recognize it as a constructive idea—instead, he interpreted it as criticizing his suggestion and felt frustrated.

Moreover, Alan found it challenging to adjust his communication style when collaborating. While he preferred clear, direct instructions, others in the committee discussed ideas more informally and open-endedly, expecting members to grasp suggestions without explicit statements. Alan's efforts to share his views often came across as abrupt or irrelevant, making some members uncomfortable. Unaware of these unspoken social dynamics, Alan became increasingly stressed, feeling misunderstood and isolated.

As a result, committee work that required "joint attention of behavior"—the ability to collaborate and respond to others' actions—became overwhelming for Alan. His struggle to navigate the social intricacies of group work on the HOA committee left him feeling sidelined despite his eagerness to contribute.

Struggles with Social Information

The most overarching challenge is understanding, comprehending, analyzing, synthesizing, evaluating, and differentiating social information in the environment. Adults with ASD often struggle to process the complex unspoken rules that govern social interactions. This difficulty is not a matter of intelligence but rather a difference in how information is perceived and processed. Social situations that seem simple to neurotypical individuals—like knowing when to join a conversation, how to interpret sarcasm, or understanding the nuances of a workplace hierarchy—can be deeply confusing for autistic adults. This can produce feelings of anxiety, self-doubt, or frustration as they navigate environments where social understanding is essential.

All of these core issues affect adults with ASD in broad and varied ways, touching nearly every aspect of their lives. The challenges of understanding others' mental states, reacting emotionally in a socially expected manner, planning and organizing tasks, interpreting nonverbal cues like gaze, working cooperatively, and processing social information can make navigating both personal and professional environments difficult. Yet it's essential to remember that these challenges are not insurmountable! With support, understanding, and adaptive strategies, autistic adults can build resilience, find success, and lead fulfilling lives. The key lies in recognizing that these core issues are not insurmountable barriers— they're areas where targeted interventions and accommodations can make a difference.

Understanding Resilience and Autism

Resilience refers to the ability to adapt to changes and grow despite adversity. It involves developing coping strategies, regulating emotions, and honing problem-solving skills that enable individuals to face challenges with confidence and effectiveness. In our work, resilience is viewed not as an innate quality but as a collection of skills and behaviors that can be cultivated over time. Our research underscores that the experiences, mindset, and support systems available to individuals shape resilience. While resilience is essential for everyone, it holds particular importance for autistic adults, who often encounter unique and complex challenges in their daily lives.

For neurotypical individuals, resilience may involve dealing with setbacks, managing stress, and maintaining a positive outlook in the face of adversity. However, for autistic adults, resilience demands even greater effort due to the unique obstacles they face—in addition to the common challenges everyone encounters, individuals with ASD must learn to navigate a world that may not always accommodate their needs or recognize the difficulties they're experiencing. For this reason, resilience for autistic individuals often requires having specialized strategies, enhanced self-awareness, and an ability to adapt in ways that address their specific challenges.

Resilience is *not* about eliminating difficulties, it's about learning to face and adapt to them! One of the core principles of our work is that resilience involves the ability to find meaning and purpose in adversity. For adults with ASD, this means learning to manage and overcome the social, emotional, and sensory challenges they encounter daily. It requires developing coping strategies that will assist them in regulating their emotions, managing stress, and navigating complex social environments more effectively. Such strategies may include implementing mindfulness practices to remain grounded in overwhelming situations, using sensory tools like noise-canceling headphones, or engaging in cognitive behavioral therapy (CBT) to reframe negative thoughts and enhance emotional resilience.

Our previous work has emphasized the importance of developing problem-solving skills as a vital aspect of resilience. For autistic adults, problem-solving often necessitates having a structured approach to tackle specific challenges. This might involve breaking tasks into manageable steps, establishing routines that offer stability and predictability, or learning to communicate needs effectively in social or professional contexts. By honing these problem-solving skills, autistic adults can become more proficient at managing their challenges and discovering solutions that suit them.

Resilience also includes the ability to build and maintain supportive relationships, which presents particular challenges for many individuals with ASD. Social interactions can be confusing or stressful— interpreting body language, facial expressions, and tone of voice is not always intuitive for people on the autism spectrum. However, with appropriate support and guidance, autistic individuals can learn to manage these social challenges and foster meaningful relationships that enhance their sense of belonging and emotional well-being. In our work and in this book, we emphasize that having supportive relationships with family, friends, or professionals is fundamental to resilience. These people offer encouragement, understanding, and practical assistance during difficult times, providing a safety net that helps individuals recover from setbacks more quickly.

We also emphasize how important it is to foster a sense of personal control and autonomy as a core aspect of building resilience. Autistic adults often feel that much of their environment is beyond their control, especially in situations involving sensory overload or unclear social expectations. However, by developing skills that enhance self-regulation and provide a greater sense of control over their environment, individuals with ASD can strengthen their resilience. This may involve learning self-advocacy techniques, creating sensory-friendly spaces, or establishing routines that promote stability.

Ultimately, resilience is about more than just surviving difficult situations—it's about learning from adversity. For autistic adults, resilience offers a pathway to greater independence, personal fulfillment, and success. It allows them to face their challenges with confidence and develop the tools they need to lead a meaningful, satisfying life. Resilience is not a fixed trait but a skill that can be nurtured and developed over time, giving individuals with ASD the ability to adapt and grow in a world that may not always understand their unique challenges. Through self-awareness, problem-solving, emotional regulation, and the support of meaningful relationships, autistic adults can cultivate resilience and use it as a foundation for personal growth and long-term well-being.

This understanding of resilience provides a hopeful and empowering framework for individuals with ASD. It illustrates that although the obstacles they encounter are real and significant, those obstacles can indeed be surmounted with the right mindset and strategies. Through resilience, autistic adults can transcend merely coping with their condition to lead a life filled with opportunity, meaning, and growth.

In order for autistic adults to cultivate resilience, first they must comprehend the condition itself. ASD is not a one-size-fits-all diagnosis—it exists along a spectrum. This means that individuals with ASD experience the condition differently depending on their specific symptoms and the severity of those symptoms. For instance, some autistic adults may excel in intellectual pursuits or demonstrate exceptional talents in areas such as mathematics, technology, or the arts. However, those same individuals might find social interactions to be overwhelming or confusing. Conversely, others may struggle more significantly with executive function skills, such as time management, organization, or prioritization, which can impact their daily living and ability to achieve personal or professional goals.

Understanding ASD in adulthood also requires recognizing that the challenges associated with the condition can evolve. While many individuals with ASD receive early intervention and support as children, transitioning to adulthood often brings new challenges. For instance, maintaining employment, living independently, and forming mature relationships can feel daunting. Social expectations become more complex in adulthood and the need for independence increases, which may exacerbate difficulties related to communication, emotional regulation, and sensory sensitivities.

Autistic adults may also experience heightened levels of stress and anxiety, which can further complicate social interactions and relationships. However, with appropriate support systems and access to resources like therapy, training, and social skills coaching, autistic adults can navigate these social challenges more effectively, thereby improving their quality of life and increasing their chances of success.

The Importance of a Resilient Mindset

A resilient mindset is crucial for individuals with ASD to overcome the unique hurdles they face. It involves cultivating specific attitudes and behaviors that help individuals cope with adversity rather than being overwhelmed. This mindset fosters personal control, emotional regulation, and the ability to confront challenges head-on. For autistic adults, developing a resilient mindset requires both internal and external support systems: therapy, coaching, and encouragement from family members and friends, for example.

One of the critical components of resilience is the ability to reframe challenges as opportunities for growth. Consider Sophie, a 35-year-old woman with ASD who struggled with social anxiety at work. Sophie often found team meetings and social events overwhelming, and her initial response was to avoid these situations. That only increased her sense of isolation and anxiety. However, with the guidance of a therapist, Sophie began to view these challenges differently. She started to see team meetings not as a source of fear but as an opportunity to practice her social skills in a structured environment.

Together with her therapist, Sophie developed strategies to manage her anxiety, such as using deep breathing exercises before meetings and preparing conversation topics in advance. Sophie's confidence grew as she gradually exposed herself to these situations and applied her coping techniques. Over time, she found that each social interaction, though uncomfortable at first, contributed to her personal growth. Through this process, Sophie improved her ability to manage social situations and build a resilient mindset. She learned that facing her fears rather than avoiding them allowed her to thrive; through resilience, she fostered her personal and professional development.

Similarly, resilience in autistic adults can be nurtured through self-awareness and emotional regulation. For instance, Alex, a 45-year-old man with ASD, experienced emotional meltdowns triggered by sensory overload. Everyday environments like crowded shopping centers, workplaces with bright fluorescent lights, and noisy public transportation could overwhelm him, causing intense stress and frustration. These meltdowns significantly impacted his daily life and made it difficult to participate in social and professional activities.

However, Alex began to better understand his emotional triggers through resilience training and self-awareness. Working with a therapist, he learned to identify early signs of sensory overload, such as increased heart rate or irritability. With this awareness, Alex adopted various coping mechanisms to mitigate the effects of overwhelming stimuli: when stressed, he practiced deep-breathing exercises or used noise-canceling headphones to block out disruptive sounds, and he sought out quieter, dimly lit spaces in his environment where he could regain composure.

With consistent practice and support, Alex became more adept at managing his emotional responses. He learned to regulate his reactions and avoid full-blown meltdowns, empowering himself to face previously intolerable situations with greater confidence. This increased his emotional control and enhanced his sense of independence and resilience, which in turn allowed him to navigate environments that had once felt unmanageable. Over time, Alex's ability to implement these strategies contributed to a more fulfilling and less stressful life. His case certainly underscores the importance of resilience for autistic adults!

A resilient mindset also involves setting realistic goals and taking ownership of one's personal development. Many adults with ASD experience difficulties with executive function, such as organizing tasks, managing time, or setting priorities, but they can develop a sense of accomplishment and greater confidence in their abilities by breaking larger tasks into smaller, more manageable steps. This incremental progress serves to reinforce a positive self-image as well as resilience, both of which are crucial for long-term success.

Harmful Scripts and Their Impact on Resilience

One of the most significant barriers to resilience in autistic adults is the presence of "negative scripts"—deeply ingrained beliefs or patterns of thought that undermine self-confidence and reinforce feelings of inadequacy. (We'll discuss this in greater depth in the next chapter.) These harmful scripts often develop early in life due to experiences of failure, rejection, or misunderstanding. For instance, an autistic adult may believe they're incapable of forming meaningful relationships because of past difficulties with social interactions. This belief becomes a negative script that influences future behavior, leading to an avoidance of social situations and a deepening sense of isolation.

Harmful scripts are particularly damaging because they're self-fulfilling. When individuals with ASD internalize the belief that they're socially inept or incapable, they may approach social interactions with heightened anxiety. This anxiety in turn affects their behavior, reinforcing adverse outcomes and perpetuating the harmful script. Breaking free from these scripts is essential for fostering a resilient mindset.

Marsha, a 36-year-old woman with ASD, has struggled her entire life to form friendships. She faced social rejection throughout childhood, leading her to believe she was unlikable and incapable of sustaining relationships. This negative narrative followed her into adulthood, where she avoided social interactions to prevent further rejection. However, through resilience training, Marsha learned to challenge this negative narrative. She realized that her past experiences need not determine her future and that with practice, she could improve her social skills and develop meaningful connections. By gradually confronting her fears and embracing new opportunities for social interactions, Marsha was able to rewrite her once-negative narrative and build resilience.

Challenging harmful scripts requires self-reflection and the willingness to question long-standing beliefs. For autistic adults, this process is often facilitated by therapists or coaches who help individuals identify the harmful scripts that are holding them back. Cognitive behavioral techniques—reframing negative thoughts, practicing mindfulness, etc.—are powerful tools for breaking these destructive patterns. As individuals begin to replace harmful scripts with more positive, realistic beliefs, they lay the groundwork for a resilient mindset that can propel them forward.

Coping, Personal Control, and Empowerment

Another critical aspect of resilience is the sense of personal control, the belief that we have the power to influence our own life and make meaningful changes. For autistic adults, this sense of control can sometimes feel elusive, especially when they're faced with challenges that seem beyond their ability to manage. However, developing a resilient mindset involves recognizing that while certain aspects of life may be uncontrollable, there are always areas where personal agency can be exercised.

Jake, a 30-year-old man diagnosed with ASD, worked in a high-pressure office environment. Jake was similar to Alex—he often felt overwhelmed by the fast pace of his job and the constant sensory overload from the bright lights and loud noises. Initially, Jake felt powerless to handle these challenges, believing the environment was just too overwhelming. However, with the help of a therapist, he learned to focus on what he *could* control, and he requested accommodations from his employer, such as wearing noise-canceling headphones and adjusting the lighting in his workspace.

By taking proactive steps to manage his environment, Jake regained a sense of control and improved his ability to cope with the demands of his job. Personal control is not about eliminating all challenges but finding ways to adapt and respond to them. This mindset shift is significant for autistic adults as it encourages them to view obstacles as opportunities for problem-solving and personal development.

Moreover, self-advocacy reinforces personal control. Autistic adults who learn to advocate for themselves in social, professional, and individual settings are more likely to feel empowered and resilient. Self-advocacy involves understanding our strengths and challenges and effectively communicating those needs to others to foster cooperation and support, and for adults with ASD, developing self-advocacy skills allows them to take control of their lives and create environments where they can thrive.

Camouflaging—also known as masking—is a common coping strategy employed by autistic adults to navigate social environments that can feel overwhelming or confusing. This strategy involves consciously or unconsciously mimicking neurotypical social behaviors to fit in, such as maintaining eye contact, controlling facial expressions, or adjusting body language to appear more socially "typical." While camouflaging can help adults with ASD avoid negative judgments or social rejection, it can also be mentally and emotionally exhausting and lead to increased anxiety, stress, and burnout. Additionally, since most autistic adults struggle to understand nonverbal communication, it can be very challenging for them to mimic these forms of communication in a genuine and effective manner.

Joan, a 42-year-old woman with ASD, had spent most of her adult life perfecting the art of masking. She meticulously studied her colleagues' behaviors at work, mimicking their facial expressions and maintaining eye contact even though it made her uncomfortable. She forced herself to engage in small talk during lunch breaks and suppressed her natural tendency to withdraw whenever conversations became overwhelming. To most outsiders, Joan seemed at ease; it looked like she was blending in with the social flow of the office.

But this constant effort to mask her true self came at a significant cost, and over time, Joan began to feel increasingly exhausted. The energy she was expending just trying to fit in left her emotionally drained and anxious. Not knowing if she had selected the most important behaviors to mimic or if she was coming across as authentic was adding to her stress. She found it difficult to relax at home—she constantly replayed social interactions in her mind, worrying about whether she had said or done the "right" things. Her sleep became disrupted and she started to feel physically unwell.

Eventually, Joan reached a point of burnout: the pressure to conform became unbearable. She felt detached from her true self, isolated, and emotionally overwhelmed. She realized that although her coping strategy of masking had initially been helpful, it had also led her to severe burnout. That realization forced her to rethink how she navigated social situations.

Like Joan, we also do not believe that masking is an effective long-term strategy. Instead, our research emphasizes the significance of cultivating a socially resilient mindset to form the foundation for effective coping strategies. Social resilience for autistic adults involves not only managing social demands but also embracing and understanding their own neurodiverse identities. This mindset enables individuals to develop adaptive skills without feeling compelled to fully conform to neurotypical norms. Rather than employing ineffective techniques such as masking, adults with ASD can cultivate self advocacy skills, establish boundaries in social situations, and seek out environments that better accommodate their needs. This approach alleviates the cognitive and emotional burden associated with masking and empowers individuals to thrive more authentically.

As discussed in this book, effective coping strategies include seeking external support systems like support groups or therapy, settings where autistic adults can share their experiences without fear of judgment. Additionally, mindfulness and self-regulation techniques help manage sensory overload and social stress—for example, recognizing personal limits and using tools like scheduled breaks or noise canceling headphones in stressful environments can help prevent burnout. By combining social resilience with these adaptive coping strategies, adults with ASD can better navigate social situations while maintaining their mental health and well-being.

The Path Forward: Building Resilience

The power of resilience lies in its ability to transform challenges into opportunities for growth. For autistic adults, cultivating a resilient mindset is essential for navigating the complexities of their social, professional, and personal lives. Through resilience, individuals with ASD can better understand their unique challenges and develop the necessary tools to overcome them. The primary goal of this book is to provide adults with ASD—and their families, therapists, and caregivers—with strategies for fostering resilience, helping them thrive in a world that may not always accommodate their needs.

These pages are designed to teach adults with ASD that resilience is not an inherent trait but a skill that can be developed and strengthened through practice. It's essential to appreciate that resilience does *not* mean that we should eliminate difficulties or shield ourselves from adversity. Instead, resilience involves facing challenges with confidence, adaptability, and hope; resilience means using those experiences to foster personal growth and a sense of control over our own lives. By learning to build resilience, autistic adults can not only cope with the everyday obstacles they encounter, they can also flourish in their personal and professional lives.

A key part of building resilience involves understanding the importance of relationships with family and loved ones. Support from close friends and family is crucial for fostering emotional resilience. For autistic adults, openly communicating about their needs, boundaries, and challenges allows loved ones to provide meaningful support, creating a safe and accepting space where they can explore their potential. Together, individuals and their families can cultivate mutual understanding and replace harmful beliefs with empowering ones. Cognitive behavioral techniques and mindfulness practices assist individuals in managing stress, staying present, and strengthening relationships.

This book also highlights emotional regulation as being essential for resilience. Although adults with ASD often face heightened stress from sensory overload or social interactions, practical strategies like engaging in deep breathing and creating sensory-friendly environments help them manage their emotions and build their confidence. Social competence is another focus—in these chapters, we'll address the challenges involved in interpreting social cues and forming relationships. With step-by-step guidance, readers will learn to improve their social skills, navigate interactions more successfully, and create meaningful connections in various environments.

Five Key Takeaways

Resilience is Essential for Adults with ASD

Resilience is a vital skill that enables individuals with ASD to adapt, persevere, and flourish despite social, emotional, and sensory challenges. Resilience is not inherent but rather a learned ability that can be developed through self-awareness, coping strategies, and external support systems.

Harmful Scripts Can Hinder Growth

Many autistic adults struggle with "negative scripts"—deeply ingrained beliefs that reinforce self doubt and avoidance behaviors. Challenging and reframing these narratives is critical to developing confidence, self-efficacy, and resilience.

Personal Control Leads to Empowerment

A strong sense of personal control aids individuals with ASD in managing challenges more effectively.

Although some environmental factors may be beyond their control, concentrating on manageable aspects—like self-advocacy, emotional regulation, and structured routines—can promote greater independence and success.

Social and Emotional Challenges Require Adaptive Strategies

Challenges with theory of mind, joint attention of emotions, interpreting social cues, and joint attention of behavior (to name a few) can make personal and professional interactions difficult. Implementing targeted strategies—such as structured communication, mindfulness, and sensory management techniques—can improve social competence and alleviate stress.

Support Networks and Self-Advocacy Are Crucial

Building relationships with understanding family members, therapists, and/or support groups can help autistic individuals manage daily challenges.

Learning to self-advocate, such as requesting workplace accommodations or seeking social support, strengthens resilience and fosters personal growth.

Self-Guided Activities to Overcome the Challenges of ASD

Activity 1: Identifying Personal Challenges

Through journaling, recount specific events tied to each challenge, noting their surroundings, emotional responses, and outcomes. Afterward, analyze your entries and look for common patterns or triggers (i.e., noisy environments or unclear social expectations). Recognizing these patterns will help you develop small, actionable steps for improvement—for example, maybe you'll initiate a casual conversation once a week or carry noise-canceling headphones to manage your sensory inputs. This activity promotes awareness and gradual, realistic changes that will better support your daily functioning.

Activity 2: Recognizing Harmful Scripts

Reflect on experiences you've had that may have shaped these beliefs, like past rejections or repeated failures. After you've acknowledged these sources, rewrite your narratives as scripts that are more balanced and constructive—for instance, you can reframe "I'll never succeed in my career" as "I face challenges, but with persistence and learning, I can progress." This cognitive restructuring will promote your self-compassion and growth by transforming limiting beliefs into motivational affirmations that will better reflect your reality and support your emotional well-being.

Activity 3: Building Resilience

Create a personalized coping toolbox that includes stress-relief strategies like deep breathing, taking sensory breaks, or using comforting objects. Finally, establish a SMART goal (**s**pecific, **m**easurable, **a**chievable, **r**elevant, and **t**imebound) aimed at emotional or mental growth. For example, you might commit to practicing five minutes of daily mindfulness every week. By fostering your strengths, reinforcing beneficial habits, and setting attainable goals, this activity will empower you to face adversity with greater confidence and emotional stability.

Activity 4: Enhancing Personal Control

Drawing a "circle of control" allows you to place controllable items (like daily routines or reactions) inside the circle and uncontrollable elements (such as the behavior of others) outside the circle. Focus on one manageable aspect and create a practical plan—for example, you might want to use a planner to organize your tasks. You can also practice self-advocacy by drafting a script to communicate a personal need, such as requesting a quieter workspace. These actions will give you a greater sense of agency, reduce your frustration, and encourage you to engage in thoughtful communication to improve your daily functioning and self-esteem.

Activity 5: Building Support Networks

Identify three trusted people (friends, family, and/or professionals) for support and encouragement. Selecting people to meet with, discuss a personal challenge with, or spend enjoyable time with will enable you to foster more proactive connections—reaching out cultivates emotional closeness and reduces feelings of isolation. Also, it's a good idea to engage with communities that align with your interests or experiences, such as online forums or local groups for adults with ASD. These networks create safe environments for mutual support and understanding, allowing you to feel more accepted, more empowered, and less isolated on your personal development journey.

Activity 6: Reflecting on Progress

Note at least one small achievement (perhaps you contributed to a meeting without feeling anxious) to validate your efforts and enhance your confidence. A personal journal can serve as a tool to monitor your long-term advancements in areas like resilience, sensory regulation, or overcoming negative self talk. Gradually, you'll spot trends and see where you're improving, giving you a sense of accomplishment and the motivation to persist. By recognizing your small victories and reviewing your progress, you'll enhance your self-awareness, and you'll also have more sustained momentum to achieve your personal objectives, reaffirming that even minor actions lead to significant changes.

Completing these activities will enable you to actively engage with the ideas discussed in this chapter to foster having more resilience, greater personal control, and a more empowered mindset. You can regularly revisit these activities to monitor your growth and adapt them as your needs and goals evolve. Don't be discouraged by setbacks! Instead, view them as opportunities to learn and develop more realistic goals and strategies. You will find that some of these same activities are also suggested in subsequent chapters given the close association between the different components of resilience. Strengthening one feature of our social = emotional, resilient lives often has the benefit of strengthening other features as well.

2

Rewriting Negative Scripts for Adults with ASD

The concept of "negative scripts" has long been a focus of our work. It's rooted in the understanding that deeply ingrained thought patterns can significantly affect how individuals perceive themselves and interact with the world. The term initially emerged from studies on cognitive behavioral therapy, which emphasize the power of internal narratives in shaping behavior and emotional well-being. Over the years, we've adapted this concept to tackle the unique challenges faced by individuals encountering various difficulties, including those with ASD. We recognize that the scripts they develop are often shaped by repeated experiences of misunderstanding, rejection, or frustration in navigating a neurotypical world. These scripts are not merely thoughts—they represent frameworks that influence how individuals interpret events, respond to challenges, communicate with others, and perceive their potential for growth. Recognizing and addressing these scripts is vital for fostering resilience and empowerment among those with ASD.

For autistic individuals, rewriting harmful scripts provides a transformative pathway to empowerment, growth, and resilience. In contrast, negative scripts gradually entrench feelings of inadequacy, restrict personal growth, and reinforce damaging stereotypes, creating a cycle that can be difficult to break. These narratives often stem from the real challenges associated with ASD, such as needing to interpret social cues, manage sensory overload, or navigate emotionally charged situations. The societal lack of awareness regarding these challenges can further exacerbate feelings of isolation and misunderstanding.

Harmful scripts can take various forms, including internalized beliefs like "I'll never connect with people because I can't maintain eye contact" or

"I don't belong in group settings because I'm always misunderstood." Such thoughts may lead to withdrawal from social opportunities, avoidance of new experiences, and diminished self-esteem, all of which reinforce the notion that growth is unattainable. However, by recognizing that these scripts are not absolute truths but rather interpretations shaped by external circumstances and challenges (.e.g., a noisy, crowded office), individuals with ASD can begin to liberate themselves from their constricting beliefs.

Breaking a script into manageable pieces allows individuals to start reframing it—instead of internalizing the belief as a limitation, they can view it as a problem to solve and can identify actionable steps like requesting accommodations, using sensory tools, or seeking a more supportive work environment. This process not only rewrites harmful scripts, it also builds resilience and empowers individuals with ASD to approach challenges with self-compassion and confidence. Recognizing the origins of these scripts allows them to embrace their strengths, seek accommodations, and develop strategies to thrive.

Why Individuals with ASD Are Particularly Vulnerable to Negative Scripts.

Negative scripts are deeply ingrained thought patterns that shape an individual's perceptions, behaviors, and self-esteem. For autistic adults, these harmful scripts have often developed in response to repeated experiences of misunderstanding, rejection, or frustration in a world designed for neurotypical individuals. Due to the cognitive, social, and sensory challenges associated with ASD, these scripts can become particularly entrenched, making them more difficult to change. Understanding why individuals with ASD are vulnerable to negative scripts offers insights into the importance of having targeted strategies for rewriting them.

Cognitive Processing Differences and Rigid Thinking Patterns

People with ASD often show cognitive processing styles marked by black-and-white thinking, a strong preference for structure, and challenges with cognitive flexibility. These tendencies can make it difficult to reassess and alter established thought patterns. When negative experiences happen, autistic individuals may generalize them and form rigid beliefs like "I always fail at social interactions" or "No one will ever understand me."

Luke, a 22-year-old college student diagnosed with ASD, has always excelled academically but finds social interactions overwhelming. He prefers structured environments and struggles with ambiguity—he often feels anxious in unpredictable situations. One day, Luke attempted to join a group discussion in his psychology class. He misunderstood a joke made by one of his peers and responded literally, which led to awkward laughter from the group. Feeling embarrassed, Luke concluded, "I always say the wrong thing. No one wants to talk to me."

This single negative experience reinforced his already rigid belief that he was socially inept. Despite previous instances where he had successfully engaged in conversations, he fixated on this one particular failure. Even when his professor later reassured him that misunderstandings are common and his insights were valuable, Luke dismissed the professor's encouragement, believing, "They're just saying that to be nice."

Unlike neurotypical individuals who might naturally reinterpret or reframe negative experiences over time, autistic people struggle to adjust their internal narratives due to difficulties with flexible thinking. This rigidity reinforces harmful scripts, making them feel absolute and unchangeable.

Frequent Social Misunderstandings and Rejection

Many people with ASD face social challenges due to difficulties with interpreting social cues, maintaining conversations, or following unwritten social norms. These struggles often result in repeated misunderstandings, exclusion, or even direct rejection from peers, colleagues, or authority figures. Over time, these experiences can foster harmful narratives such as "I am socially awkward and will never fit in" or "People don't like me."

Unlike neurotypical individuals—who may have a balance of both positive and negative social interactions—many autistic people accumulate more negative social experiences. Without adequate support or opportunities to develop social confidence, these repeated experiences reinforce harmful scripts that can lead to avoidance behaviors and social withdrawal.

Heightened Sensory Sensitivities

Many autistic people experience heightened sensory sensitivities, making environments like classrooms, workplaces, and public areas overwhelming. Sensory overload can lead to distress, shutdowns, or difficulty concentrating, which others may misinterpret as behavioral issues or a lack of competence.

As a result, individuals with ASD may internalize harmful beliefs like "I can't handle normal environments" or "I'm not capable of succeeding in work or school."

For instance, Morris, a 20-year-old college student with ASD, found the fluorescent lights and background chatter in his lecture hall unbearable. Despite his strong academic abilities, he struggled to focus and often left class early due to headaches and exhaustion. His professors misinterpreted his early exits as a lack of motivation, while his peers perceived him as being disinterested. Over time, Daniel began to believe that he was incapable of managing a college environment, an assumption that reinforced a cycle of self-doubt and avoidance.

Emotional Regulation Challenges

Emotional regulation challenges further compound these difficulties. Many individuals with ASD struggle to process and regulate emotions effectively, which can amplify feelings of frustration, anxiety, and self-doubt. This emotional intensity makes it easier for negative scripts to become deeply embedded, as each challenging experience reinforces an already existing narrative of inadequacy or failure.

When people offer accommodations such as sensory-friendly learning spaces and self-regulation strategies, challenge an individual's negative self-perceptions, and foster confidence in their abilities, individuals like Daniel can be empowered to navigate their environments more successfully.

Societal Expectations and Pressure to Conform

Society's often rigid expectations regarding success, communication, and social behavior marginalize neurodivergent individuals. From an early age, autistic individuals are regularly compared to neurotypical peers, receiving messages that they need to "try harder" to fit in. This societal conditioning perpetuates damaging beliefs such as "I'm not good enough," "I'm broken," or "I'll never be successful."

The workplace further reinforces these narratives. Many professional environments prioritize traits such as adaptability, effective verbal communication, and high levels of social engagement— qualities that can be challenging for individuals on the spectrum. Without adequate accommodations and understanding from others, individuals with ASD may face professional setbacks that deepen feelings of inadequacy and failure, making it even more difficult to challenge and replace harmful scripts.

For instance, Susan, a 30-year-old skilled software developer with ASD, faced challenges at her job due to the constant expectation to engage in small talk and attend team meetings. Although her work was of high quality, she received negative feedback about her "lack of collaboration." Her employer failed to understand her need for structured communication rather than spontaneous discussions. Ultimately, feeling undervalued and overwhelmed, Susan decided to resign, reinforcing her belief that she could not achieve professional success.

Repetition and the Deep-Rooted Nature of Negative Scripts

Because people with ASD often thrive on routine and predictability, they may unknowingly reinforce harmful scripts through repetition—the brain strengthens neural pathways associated with frequently repeated thoughts, making negative scripts feel more "natural" over time. The longer a negative script is reinforced, the harder it becomes to challenge and replace that script.

For example, if an autistic person has had several unsuccessful social interactions, they may avoid future opportunities due to fear of rejection. This avoidance behavior prevents them from gathering new, positive experiences that could counteract their existing harmful script. Over time, this cycle deepens the belief that they are socially incapable, reinforcing their internal negative narrative.

The Role of Practice and Repetition

Reframing harmful scripts isn't an overnight transformation—integrating new narratives into daily life requires deliberate effort, consistent practice, repetition, and an ability to deal with setbacks. Altering deeply rooted thought patterns takes time, but the good news is that each step forward bolsters the belief that positive change is indeed attainable. The process involves actively engaging with these new perspectives, applying them in real-world situations, and revisiting them whenever challenges arise.

One effective way to integrate reframed narratives is through role-playing scenarios with a supportive therapist, mentor, or trusted friend. These practice sessions simulate real-life situations, allowing individuals to rehearse alternative responses and behaviors in a safe and encouraging environment. For example, someone who struggles with receiving workplace feedback may role-play how to stay calm, ask clarifying questions, and view feedback as a tool for growth rather than criticism. This rehearsal helps replace defensive reactions with constructive engagement, making navigating similar real-life situations less challenging.

Repetition is essential for making these new scripts feel natural. Regular practice—whether through structured exercises, journaling, or guided conversations—reinforces the revised narratives. For instance, a person might begin each day by writing affirmations that reflect their evolving beliefs, such as "I am capable of adapting to new challenges" or "My unique strengths add value to my team." Over time, these repeated affirmations can transform internal dialogue and change behaviors, enhancing confidence and self-esteem.

Building comfort and confidence through practice is an empowering process. Each small success—navigating a difficult conversation, managing sensory overload, handling constructive criticism—reinforces the belief that growth is possible. This iterative approach fosters resilience and builds a sense of agency, helping individuals take ownership of their personal development and feel more empowered on a daily basis.

The Impact of Societal Expectations on Negative Scripts

Societal expectations play a critical role in shaping how individuals view themselves. Unfortunately, for autistic adults, these expectations often reinforce damaging internal narratives. From an early age, neurodivergent individuals are frequently compared to neurotypical peers, leading to a persistent feeling of inadequacy. Expectations surrounding social interactions, career achievements, and success in relationships create a rigid framework that may not consider the unique strengths or challenges of those with ASD. As a consequence, when these societal benchmarks aren't met, individuals internalize feelings of inadequacy, in the process reinforcing negative scripts that suggest they are incapable, unworthy, or fundamentally different in an undesirable way.

Workplace culture often magnifies these detrimental narratives. Many professional environments prioritize traits such as adaptability, strong verbal communication, and high levels of social engagement—qualities that can be particularly challenging for individuals on the spectrum. Those who struggle with office politics, small talk, or sensory overload may find themselves overlooked for promotions or subjected to unfounded assumptions about their capabilities. The pressure to conform to neurotypical expectations can lead to exhaustion and self-doubt, reinforcing the notion that certain people aren't suited for success in professional settings. Without understanding and accommodations, the workplace becomes yet another arena where damaging scripts take root.

Traditional success metrics like financial status, relationship milestones, and social popularity also shape these narratives. Autistic individuals may take longer to achieve career stability; they may form relationships in their own ways or prefer solitary pursuits that don't align with mainstream definitions of social success. This disconnect creates the misconception that they're failing in life when in reality, they're simply functioning within a different framework. Challenging these external pressures begins with redefining success to honor neurodivergent strengths. Instead of evaluating their achievements by neurotypical standards, individuals can focus on personal fulfillment, mastery of their interests, and making meaningful contributions to their communities.

Rewriting these scripts requires actively resisting societal conditioning. One way to do this is by seeking out and celebrating neurodivergent role models who have redefined success on their terms. Engaging in communities that embrace differences and support self-acceptance also fosters a sense of belonging. Advocating for accommodations in workplaces and social settings can create environments that nurture confidence rather than diminish it. The more individuals challenge these external pressures, the more they can rewrite their internal narratives to reflect their worth, capabilities, and potential.

The Role of Self-Compassion in Rewriting Scripts

Self-compassion is a vital yet often overlooked aspect of reshaping harmful narratives. Many autistic adults have spent years internalizing messages of inadequacy, resulting in cycles of self-criticism and doubt. Self-compassion disrupts this cycle and enables individuals to acknowledge their challenges without perceiving them as personal failures. Unlike self-esteem—which often depends on success as society defines it—self-compassion offers a consistent foundation of acceptance regardless of external outcomes. By embracing self-compassion, individuals can confront damaging narratives and replace them with stories that foster resilience and self-worth.

One of the most effective ways to cultivate self-compassion is through mindfulness. Mindfulness allows us to observe our thoughts and emotions without immediate judgment. Instead of reacting to negative self-perceptions with further self-criticism, mindfulness encourages a more neutral and understanding approach. For example, when facing a difficult social situation, instead of thinking, "I always say the wrong thing; I'm socially inept," an individual practicing mindfulness might acknowledge, "I feel uncomfortable in social situations, and that's okay. Many people struggle with this." This shift from self-judgment to self-acceptance is a key step in rewriting scripts.

Practical exercises reinforce self-compassion. One powerful method involves us writing a letter to ourselves from the perspective of a supportive friend. This exercise enables us to view our struggles with kindness rather than criticism. Another effective strategy is to develop self-compassionate affirmations, like telling ourselves "I am doing my best, and that's enough" or "I deserve kindness even when I make mistakes." Repeating and applying these affirmations on a daily basis can gradually replace self-critical thoughts with a more supportive internal dialogue.

Practicing self-compassion also involves setting boundaries that safeguard mental and emotional well-being. Many autistic adults often push themselves beyond their limits to meet neurotypical expectations, and that can lead to burnout and frustration. Learning to say no, requesting accommodations, and recognizing when to take a break are acts of self-kindness that enhance healthier self-perception. When individuals treat themselves with the same patience and understanding that they would extend to a friend, they foster a mental environment where positive scripts can thrive.

Breaking the Cycle: Strategies for Long-Term Script Transformation

Rewriting harmful scripts isn't a one-time event—it's an ongoing process that requires intentional effort. Over the years, thought patterns become deeply ingrained in the brain, and breaking this cycle demands awareness and consistent reinforcement. That said, understanding the neurological basis of these scripts can help individuals approach the process with patience and strategy.

In a nutshell, the brain relies on neural pathways that have been strengthened over time, meaning that negative self-perceptions often become automatic responses. However, through neuroplasticity—the brain's ability to form new connections—positive scripts can gradually replace these harmful patterns. Visualization techniques are a powerful tool for reinforcing new narratives. We can strengthen alternative neural pathways by mentally rehearsing positive outcomes and self-affirming beliefs. For example, someone who fears social rejection can visualize a successful conversation where they feel understood and valued. Over time, this practice makes it easier to gain confidence in real-life situations. Journaling is another effective method for breaking the cycle—writing down harmful scripts and actively rewriting them as constructive statements fosters a more balanced perspective. Tracking progress in a journal also serves as a reminder of growth, reinforcing the belief that change is indeed possible.

Technology can likewise be invaluable in script transformation. Reminder apps that prompt individuals to practice positive affirmations, AI-driven coaching tools that offer encouragement, and journaling apps that track patterns over time help maintain consistency. Utilizing these resources provides structured reinforcement that makes new scripts feel more natural and better integrated into daily life.

Ultimately, breaking the cycle requires persistence and self-awareness. Whenever an individual consciously replaces a negative thought with a more empowering one, they reinforce a healthier internal narrative. Over time, these small, deliberate actions accumulate, leading to profound and lasting transformation.

Navigating Relationships While Reframing Scripts

Harmful scripts can significantly impact relationships and influence how autistic individuals navigate friendships, romantic partnerships, and professional interactions—when negative narratives imply that they're unworthy of connection or incapable of forming meaningful relationships, then social engagement transforms into a source of anxiety instead of fulfillment. This often leads to withdrawal, further reinforcing the isolation that these scripts predict. Reframing these beliefs requires a shift in both perception and approach.

One of the first steps in reframing social scripts is recognizing personal strengths within relationships. Many individuals with ASD excel at having deep, meaningful conversations, being loyal, and being honest—qualities that form the foundation of strong connections. By shifting the focus from perceived deficits to these strengths, autistic adults can foster a more positive self-image in social interactions. Instead of thinking "I can't maintain friendships because I don't understand social cues," an individual can reframe that as "I build strong friendships with people who value direct and honest communication."

Setting boundaries is essential for maintaining healthy relationships and safeguarding mental well-being. Recognizing personal limits and advocating for individual needs also helps prevent social exhaustion and feelings of resentment. Clearly communicating preferences (e.g., limiting group gatherings or requesting direct communication) ensures that relationships are based on mutual understanding rather than enforced conformity.

For autistic adults, learning to advocate for themselves is vital to rewriting relationship-based scripts. Clearly communicating needs and expectations fosters healthier dynamics in friendships, romantic relationships, and workplaces, and seeking supportive communities that value neurodivergent traits instead of viewing them as deficits can greatly influence self-perception. By developing relationships in environments that promote authenticity, individuals with ASD can replace unhelpful scripts with scripts that reflect their worth and capacity for meaningful connection.

Luke, the college student described earlier who struggled with social interactions, received assistance from his therapist—together, they challenged his rigid thoughts with cognitive behavioral techniques. By analyzing past interactions and identifying moments of success, Luke gradually began to realize that not every conversation was a failure. However, his progress was slow, emphasizing the deeply rooted nature of cognitive rigidity in individuals with ASD.

Overcoming Challenges and Setbacks

Rewriting harmful scripts is not a linear process—setbacks are an inevitable part of the journey. Although discouraging at times, these challenges provide essential opportunities for learning and growth. Anticipating difficulties and recognizing that they do *not* signify failure but rather serve as steps in a larger process can help individuals maintain a resilient and adaptive mindset. By understanding the unique cognitive and social aspects of ASD, individuals can reframe harmful scripts into empowering narratives that align with their particular strengths, which in turn fosters resilience. For example, rather than viewing difficulty in maintaining eye contact as a flaw, it can be seen as a sign of reflection. Similarly, individuals can approach group settings by focusing on structured environments or by finding like-minded communities where understanding and acceptance are prioritized. By reshaping these narratives, autistic adults can build self-efficacy, recognize their inherent value, and authentically embrace their potential for growth and success.

Progress is not about achieving perfection but about embracing persistence and having the willingness to try again. Whenever a new script or strategy doesn't work as planned, it's an opportunity to reflect on what happened and why. For instance, if a social interaction feels unsuccessful despite someone having prepared for the scenario, they can analyze the experience (often with the input of a therapist or coach) to identify specific difficulties, such as unclear cues or unexpected responses from others. This reflection allows individuals to make targeted adjustments, such as refining their communication skills or seeking additional support. Every setback provides valuable insight that informs the next attempt and creates a cycle of improvement.

Reframing setbacks as temporary challenges instead of permanent failures fosters self-compassion and resilience. It's crucial to celebrate small victories no matter how minor they may seem, as these moments demonstrate progress and capability. For instance, when someone acknowledges having had the courage to start a conversation or step outside their comfort zone, that strengthens their commitment to growth. These achievements build confidence and serve as reminders that even imperfect efforts are steps in the right direction.

Again, external support is crucial in navigating challenges. A therapist, mentor, or supportive friend can provide encouragement, perspective, and practical advice during difficult times. Their guidance normalizes setbacks as being part of growth; this kind of support fosters a shared understanding and alleviates feelings of isolation.

By embracing setbacks as learning opportunities and focusing on persistence rather than perfection, individuals build the resilience they need to rewrite harmful scripts and achieve meaningful, lasting change. This mindset reinforces the fact that growth is an ongoing process and that every effort contributes to a stronger, more empowered self.

Recognizing Harmful Scripts

Rewriting harmful scripts begins with understanding their origins and identifying the events or experiences that shaped those narratives. For individuals with ASD, these scripts often stem not from personal inadequacies but from a fundamental mismatch between their unique strengths and the expectations of a society primarily designed for neurotypical individuals. Recognizing this misalignment is the first step toward change as it shifts the perspective from self-blame to self-empowerment.

Harmful scripts may be rooted in early life experiences, such as being misunderstood by peers, facing academic challenges without appropriate support, or struggling with sensory sensitivities in environments that weren't accommodating. These experiences can reinforce beliefs like "I'm not capable" or "I'll never be good at making friends." Such thoughts become internalized over time and shape how individuals perceive themselves and approach challenges in adulthood. However, recognizing that these beliefs arise from external circumstances rather than intrinsic flaws opens the door to change.

Reflection—often guided by a therapist, coach, or trusted mentor—is essential for uncovering and addressing entrenched patterns. A therapist might encourage someone to explore the origins of a recurring belief, such as "I can't handle sensory overload in a workplace." Together, they could analyze specific experiences that contributed to this thought, like struggling in a noisy, open-plan office. By breaking the script into manageable pieces, they can begin reframing it. Instead of internalizing the belief as a limitation, the individual can view it as a problem to solve, identifying actionable steps such as requesting accommodations, using sensory tools, or finding a more supportive work environment.

This process not only rewrites harmful scripts but also builds resilience, empowering autistic individuals to tackle challenges with self-compassion and confidence. Recognizing the origins of these scripts enables them to embrace their strengths, seek accommodations, and develop strategies to thrive.

Reframing Scripts as Empowering Narratives

Once harmful scripts are recognized, they can be reframed in terms of constructive and realistic alternatives that can transform negative self-perceptions into empowering narratives. This reframing process does not ignore past struggles or challenges but instead incorporates them as part of a journey toward growth and self-understanding. Acknowledging strengths and potential is a central part of this approach. For instance, a pervasive belief like "I always say the wrong thing in conversations" can be reshaped into "Conversations are challenging for me, but I can improve with preparation and practice." This new narrative shifts the focus from perceived failure to personal growth and progress.

Reframing scripts also requires individuals to recognize and celebrate their unique strengths, whether that be attention to detail, problem-solving abilities, or specialized interests. They can build a more confident and authentic self-image by emphasizing these positive attributes. For example, someone who excels at structured, task-oriented settings may redefine success by pursuing roles that align with these strengths rather than conforming to societal expectations and prioritizing unstructured social interactions.

Embracing authenticity is another crucial aspect of reframing. It involves letting go of the need to meet external definitions of success and instead focusing on values and goals that genuinely resonate. Whether it's thriving in one-on-one conversations, excelling in a niche field, or finding joy in routine, when individuals reframe scripts, they foster a sense of empowerment and alignment with their true selves.

Addressing Sensory and Emotional Needs

Addressing sensory and emotional needs is a foundational step in fostering resilience and reframing harmful scripts for autistic individuals. Many negative narratives originate from the stress and discomfort caused by navigating environments that are not attuned to their sensory and emotional experiences—over time, these challenges can contribute to feelings of inadequacy or frustration. But the good news is that individuals can create a stable foundation for personal growth and intentional engagement with their goals by prioritizing their sensory and emotional well-being. Sensory-friendly strategies play a crucial role in reducing stress and building confidence! For example, noise-canceling headphones in noisy environments can help mitigate auditory overload, while dimming lights or wearing tinted glasses may alleviate visual sensitivities.

Establishing designated calming spaces at home, work, or school offers a retreat whenever sensory inputs become overwhelming. As we saw with 30-year-old Jake, these minor but impactful adjustments allow individuals to feel more in control of their surroundings, creating a sense of safety and predictability. When the sensory environment is managed effectively, it becomes easier to focus on personal development and actively reframe unhelpful scripts.

Addressing emotional needs is equally important. Emotional regulation strategies like mindfulness exercises or deep breathing techniques can help individuals navigate intense emotions that may arise during challenging situations. Understanding and labeling emotions—whether those are anxiety, frustration, excitement, or something else—enhances self-awareness and provides clarity about what triggers these responses. This understanding makes it easier to develop tailored coping mechanisms. For instance, practicing progressive muscle relaxation can help release tension during moments of sensory or emotional stress.

Integrating these sensory and emotional strategies into daily life fosters intentional engagement with reframed narratives. When individuals feel emotionally and physically balanced, they're better equipped to challenge harmful scripts and replace them with constructive alternatives. Addressing sensory and emotional needs isn't just a practical consideration—it's a transformative process that empowers autistic adults to thrive and align their inner experiences with their external goals.

Harnessing Strengths and Interests

Harnessing strengths and interests is a transformative strategy for individuals with ASD to counter harmful scripts and cultivate a positive sense of self. People often overlook their personal strengths—whether those are in areas requiring precision, creativity, analytical thinking, or deep focus—in favor of traditional benchmarks of success. Yet these strengths are valuable assets that can provide a sense of accomplishment and pride when they're recognized and nurtured. This effectively challenges negative narratives and fosters growth.

Our work on "islands of competence" builds directly upon this idea: the concept emphasizes the importance of identifying and celebrating areas where individuals naturally excel. Introduced as a strengths-based framework, honing in on islands of competence shifts the focus from deficits to abilities, creating opportunities to bolster self-esteem and resilience. For autistic individuals, these islands of competence might manifest in areas such as mathematics, art, technology, or specialized hobbies where their unique perspectives and skills shine.

In our work, we've seen how recognizing these islands of competence can transform detrimental scripts. For instance, someone who internalized the belief of "I'll never be successful because I struggle with social interactions" might discover a remarkable talent in graphic design or coding. By channeling their energy into their strengths, individuals can build confidence and reframe their narrative as something empowering, such as "I can excel in my field by leveraging my skills and creating a fulfilling niche."

For autistic adults, their strengths often align with specific interests, and those interests can become powerful tools for connection and self-expression. For example, a person passionate about technology might excel in areas like coding, graphic design, or robotics. These skills provide opportunities for achievement and create a natural pathway for building relationships in environments that feel authentic and comfortable. Engaging in communities focused on shared interests—i.e., technology clubs and online forums—allows people to reframe negative thoughts like "I don't belong" into empowering alternatives like "I thrive in spaces where I share my passions with others."

Celebrating strengths also shifts the focus away from limitations, reinforcing a narrative of capability and potential. For example, a person with a remarkable memory might use that skill to excel at research, history, or data analysis. Recognizing these abilities transforms harmful thoughts like "I'm not good enough" into "I add unique value through my specialized skills." This process boosts self-esteem and helps individuals identify opportunities that align with their natural talents.

Focusing on strengths also fosters authenticity. Rather than striving to meet societal expectations that may feel uncomfortable or unattainable, individuals can redefine success on their own terms. Leveraging their strengths and interests empowers people to navigate challenges with confidence, transform passion into purpose, and cultivate meaningful connections that affirm their identity and value.

Transformative Real-Life Examples

The following real-life stories exemplify the transformative potential of rewriting harmful scripts, offering hope and a roadmap for change.

Sophie's Journey to Connection

We met Sophie in the last chapter. She's a 35-year-old woman diagnosed with ASD who struggles with social anxiety. For much of her life, she felt that meaningful friendships were out of reach due to repeated experiences of rejection—her unique interests and social challenges often left her feeling isolated and unrelatable. This harmful narrative of "I am unrelatable and will never have close friends" served as a barrier to forming connections, as she avoided social situations out of fear of being rejected and misunderstood. Her self-perception was further reinforced when well-meaning peers expressed confusion about her intense focus on niche topics.

With the help of therapy, Sophie began unpacking these experiences and identifying moments where her passions had sparked genuine connections. She remembered times when her knowledge of ancient history or indie films had intrigued others and led to engaging conversations. Her therapist encouraged her to view these moments as evidence that her interests could be bridges rather than barriers. Together, they refuted her harmful script and transformed it into an empowering one: "I can form meaningful friendships by sharing my passions with people who appreciate them."

Sophie took deliberate steps to explore this new narrative by seeking spaces where shared interests were celebrated, such as hobby groups and online forums. She joined a book club focused on historical fiction and discovered a welcoming community of like-minded individuals. Over time, Sophie formed friendships rooted in mutual respect and shared enthusiasm. These relationships provided positive reinforcement, challenging her previous belief that she was destined to remain isolated. Her growing confidence allowed her to approach social situations with optimism instead of fear; ultimately, she was able to replace her harmful script with a narrative of empowerment and connection.

Michael's Professional Transformation

Similarly, Michael, a talented 39-year-old graphic designer with ASD, struggled with a debilitating fear of criticism. Despite his undeniable talent, he avoided opportunities to showcase his work as he perceived any kind of feedback as being a personal attack. The harmful script of "I can't handle editicism" created a self-imposed barrier that stifled his growth. Michael's anxiety about being evaluated caused to decline freelance projects and avoid participating in team discussions, leaving him feeling stagnant and disconnected in his career.

Through coaching, Michael shifted his perspective on feedback. His coach emphasized that constructive criticism was a tool for growth, not a judgment of his worth. They worked together to reframe his harmful narrative into a constructive one, namely, "Constructive feedback helps me refine my skills and become a better designer." Michael practiced receiving feedback in low-pressure settings to solidify this new mindset, starting with peer review sessions in an online design community. Gradually, he became more comfortable with critiques and began to see how they sharpened his creative instincts.

Encouraged by his progress, Michael submitted his work to a local design competition and received thoughtful and detailed feedback from the judges. Instead of feeling defensive, he used their suggestions to refine his portfolio. This process bolstered his confidence and motivated him to take on more ambitious projects. Michael discovered that engaging with feedback improved his designs and enhanced his ability to collaborate with colleagues. Over time, he transformed his relationship with criticism—he came to view it as an opportunity rather than a threat, and his career flourished as a result.

The Power of Rewriting Negative Scripts

Both Sophie and Michael demonstrate the profound impact of rewriting harmful scripts— addressing their limiting beliefs with intentionality and support unlocked new opportunities for growth, connection, and success. Sophie overcame her fear of rejection to build meaningful relationships, while Michael reframed his anxiety about criticism into a pathway for professional growth. Their journeys highlight the resilience inherent in autistic individuals and the transformative power of intentional narrative change.

Harmful scripts are not fixed destinies but challenges that can be navigated and rewritten with courage, support, and perseverance. Sophie and Michael reshaped their lives through their efforts, proving that people can overcome deeply ingrained barriers and thrive on their own terms. Their stories stand as a testament to the human potential for growth and adaptation.

For everyone, rewriting harmful scripts begins with awareness, i.e., recognizing the beliefs and patterns that hinder personal growth. Rewriting scripts requires a willingness to challenge long-held assumptions as well as a commitment to self-reflection. Seeking support from trusted friends, mentors, or professionals can offer the encouragement and insights that are needed to navigate this transformation. Small, intentional actions such as reframing negative thoughts, practicing self-compassion, and embracing new, healthier narratives contribute to lasting change.

Individuals with ASD can indeed reclaim their story, redefine their future, and build a life aligned with their true potential!

Five Key Takeaways

1. *Understanding the Origins of Harmful Scripts*

Negative scripts are deeply ingrained thought patterns shaped by repeated experiences of feeling misunderstood, rejected, and frustrated in a neurotypical world. These scripts influence self-perception, behavior, and emotional well-being. Recognizing that these narratives are interpretations rather than absolute truths is the first step toward empowerment.

2. *Recognizing and Challenging Negative Narratives*

Identifying the specific experiences that led to harmful self-perceptions is crucial to the rewriting process. Many of these narratives stem from societal expectations and a mismatch between neurodivergent strengths and traditional social norms. By shifting the perspective from self-blame to self-empowerment, autistic individuals can begin to break the cycle of negative thinking.

3. *Reframing Negative Scripts as Empowering Narratives*

Transforming self-defeating beliefs into constructive alternatives allows individuals with ASD to embrace their strengths. For example, instead of seeing difficulty with making eye contact as a flaw, it can be reframed as a preference for reflection and a different mode of communication. This shift helps build confidence and reinforces personal growth.

4. *Knowing the Importance of Practice and Repetition*

Changing deeply embedded thought patterns requires consistent reinforcement. Techniques such as roleplaying, journaling, and practicing affirmations help solidify new narratives. Individuals can replace self-doubt with resilience and self-efficacy by actively engaging with positive scripts in their daily lives.

5. *Embracing the Role of Self-Compassion in Long-Term Change*

Practicing self-compassion is essential for overcoming negative scripts. Being mindful, practicing self affirmations, and seeking supportive communities can help individuals reframe their internal dialogue.

Instead of striving for perfection, individuals can embrace progress and learn from setbacks, in the process fostering long-term transformation and emotional well-being.

Self-Guided Activities to Rewrite Negative Scripts

Activity 1: Identifying Harmful Scripts

Reflect on how your negative thought patterns are tied to moments of frustration, anxiety, or misunderstanding, noting your inner dialogue and any assumptions you may have made, and then explore recurring beliefs such as thinking "I always fail" or "People don't get me" and seek the origins of those beliefs. (Such origins are often rooted in past experiences or reinforced through repeated situations.) Finally, categorize your scripts by context (social, sensory, work-related, etc.) so that you can spot patterns and prioritize areas for personal growth. Recognizing these harmful scripts is crucial for challenging and transforming them into healthier, more empowering perspectives.

Activity 2: Assessing Script Accuracy

Create a two-column chart and list instances that affirm and contradict each harmful belief. (For example, you can counter the thought of "I always mess up" with examples of success.) Reflecting on these beliefs will aid you in distinguishing between facts and assumptions. It will also help you have more balanced thinking, because you'll be considering whether or not you overemphasize failures while neglecting your successes. This methodical assessment will reveal that many harmful beliefs are not absolute truths but rather skewed perceptions influenced by selective memory or emotions.

Activity 3: Reframing Harmful Scripts

Transform negative self-talk into constructive narratives and rewrite harmful scripts as new, realistic, growth-focused versions. For example, you can change "I always say the wrong thing" to "I can improve my conversations through practice." The goal is not toxic positivity, but rather a believable shift that will open the door for personal improvement. Saying daily affirmations reinforces these reframed beliefs—perhaps you might say aloud, "I bring unique value through my creativity." Repeating affirmations each morning will help you internalize your positive traits and replace any automatic negativity with confidence and purpose. Over time, these affirmations can empower you to have more positive self-perceptions and healthier mental habits.

Activity 4: Leveraging Strengths and Interests

List areas where you feel competent or proud, highlighting skills or knowledge that you value. By creating a passion map, you can visually link your interests—like coding, art, or history—with opportunities where you can connect with others or contribute meaningfully. Select one interest or strength as an actionable goal, such as joining a hobby group or creating a personal project. Pursuing this goal will nurture your personal fulfillment and reinforce positive self-perceptions, demonstrating that strengths can be actively used to counterbalance harmful scripts.

Activity 5: Practicing Reframed Narratives

Role-play your reframed narratives with a trusted person to simulate scenarios like receiving feedback or initiating social interactions. Practicing these in a safe space builds confidence! You can also journal about similar situations—that will allow you to mentally rehearse your reframed narratives and envision positive outcomes. Through this process, you'll be able to clarify the steps you need to achieve your desired results. With consistent practice and visualization, these new narratives will begin to replace the old, automatic ones; you'll become more comfortable with unfamiliar behaviors and you'll internalize the idea that change is indeed possible with effort and support.

Activity 6: Managing Sensory and Emotional Needs

Start with a sensory audit: identify environmental triggers such as noise or lighting, then come up with improvements like wearing headphones or dimming lights. You can use a journal or an app to track your emotions throughout the day to uncover these patterns. Based on your insights, create a self-soothing toolkit (including items such as fidget toys or calming scents) to manage any distress proactively. Having personalized strategies immediately at hand will help you foster a better emotional balance and a more stable internal state that will support your growth. By identifying and responding to your sensory and emotional needs, you'll build a stronger foundation for self-regulation.

Activity 7: Overcoming Setbacks

Keep a Success Jar! That involves writing down your small victories on slips of paper and then putting them into a jar. Reviewing your victories during tough times will remind you of your progress and capabilities. This kind of combined reflection + celebration fosters having a mindset of perseverance; you'll learn that setbacks are temporary and that growth happens when you respond to difficulty with curiosity, not defeat.

Activity 8: Building a Supportive Network

Seek out communities that are aligned with your interests (online or in person) where you feel safe, valued, and understood. Share your reframed beliefs with trusted friends or mentors to receive feedback and encouragement. This will reinforce your progress and provide new insights. Mentors, too, are a powerful resource, as those who have overcome similar challenges can offer you their support. These kinds of interactions foster a sense of belonging and accountability for everyone and demonstrate that growth does not happen in isolation. Your supportive relationships will validate your new narratives and empower you to continue your personal development journey.

By completing these activities, you can actively identify and address your harmful scripts and build a foundation for resilience, self-confidence, and authentic personal growth.

3

Choosing the Path to Become Stress-Hardy Rather than Stressed Out

Living with ASD often involves navigating a world that feels overwhelming, unpredictable, and at times isolating. As we emphasized earlier, everyday experiences such as social interactions, sensory stimuli, or professional demands can increase stress, especially in environments that are not inherently accommodating—to name just a few, crowded spaces, sudden noises, and subtle social cues can become significant sources of anxiety. The cumulative impact of these stressors can make even ordinary activities like grocery shopping or attending a meeting feel daunting.

The path to becoming stress-*hardy* rather than stressed-*out* is not about eliminating these challenges but learning to adapt and thrive despite them. This requires developing strategies to manage stress effectively and building environments that reduce unnecessary strain. For autistic adults, both of these efforts involve cultivating resilience. Taken as a whole, it's a dynamic process that empowers individuals to face adversity with strength and flexibility. Resilience doesn't imply ignoring or suppressing stress but rather understanding it, acknowledging its impact, and using it as an opportunity to learn and grow. When individuals focus on self awareness, identify their personal strengths, and seek out supportive relationships, resilience becomes a practical tool for transforming stress into empowerment. This chapter explores building resilience through purpose, adaptability, and control, helping individuals transform stress into a source of growth and empowerment.

The Concept of Stress Hardiness

Stress hardiness refers to the qualities that allow individuals to confront life's challenges with confidence and determination instead of feeling overwhelmed or defeated. Stress hardiness means embracing a mindset that encourages resilience; this in turn empowers individuals to handle stressful situations more effectively and emerge stronger from them.

Researchers have identified three core components of stress hardiness: commitment, challenge, and personal control. Together, these elements create a framework that can change how individuals perceive and respond to stress.

Commitment

Commitment is finding purpose and meaning in daily activities, relationships, and long term goals. It stabilizes individuals, helping them stay grounded and focused even during periods of significant stress. When people feel connected to something meaningful, they're better equipped to endure challenges and navigate uncertainties. A strong commitment can form a foundation for resilience, allowing individuals to push through hardships with a determined mindset. Whether it's being dedicated to personal values, having a sense of responsibility toward loved ones, or pursuing a meaningful career, strong commitments motivate individuals to persist in the face of difficulties.

For instance, autistic adults often exhibit profound commitments to their passions, routines, and personal goals. All of that can aid them in navigating life's uncertainties. A software developer with ASD might be dedicated to their job and find comfort in the structured nature of coding and the logic inherent in problem-solving. Even when confronted with workplace challenges such as social interactions or sensory sensitivities, their dedication to their craft allows them to thrive. Similarly, an artist with ASD may encounter difficulties with networking or self-promotion but remain devoted to their creative endeavors, using art as a means of expression and fulfillment.

This connection to meaningful pursuits is a powerful anchor that provides clarity and direction when life becomes overwhelming. Commitment lays the groundwork for a positive mindset, enabling individuals to view difficulties as temporary and surmountable rather than overwhelming or permanent. Instead of seeing setbacks as failures, committed individuals perceive them as learning opportunities—they fuel their motivation to keep going. Commitment instills a sense of purpose and reinforces the belief that life is valuable and worth striving for, even during times of adversity.

An example of this is an adult with ASD who's training for a marathon—they may find comfort in the structure of their training regimen. Their commitment to running helps them maintain discipline and provides them with a sense of routine and progress. The repetitive nature of running can also create a calming effect, aiding in managing sensory sensitivities or anxiety.

Similarly, an aspiring writer with ASD who struggles with rejection but is resilient stays committed to their passion and views feedback as a tool for growth rather than as a reason to give up. Their unique perspective and intense focus on their craft enable them to develop rich, detailed stories that reflect their inner world.

This sense of meaning helps individuals focus on what truly matters, providing a source of inner strength and guiding them toward productive and fulfilling paths. Ultimately, commitment isn't just about persistence—it's about people investing in what brings them purpose and fulfillment. For autistic adults, commitment can be a vital source of stability, confidence, and self-expression, allowing them to thrive in ways that align with their strengths and values.

Challenge

Challenge is the second component of stress hardiness. It involves perceiving stressors as opportunities for growth and self-improvement rather than as threats. This perspective encourages individuals to tackle difficulties with curiosity, determination, and a proactive mindset. Rather than succumbing to fear or avoidance, those who adopt this outlook are more likely to view setbacks as a natural part of the learning and growth processes. This shift in perspective fosters greater adaptability, as individuals concentrate on problem-solving and creative thinking instead of dwelling on negative outcomes.

By viewing challenges as inevitable and even valuable parts of life, individuals can lessen the tendency to catastrophize and see hardships as insurmountable. Instead, obstacles transform into stepping stones for personal growth, teaching resilience and building confidence. This mindset improves an individual's ability to handle current stressors and prepares them to confront future challenges with greater assurance, knowing that they possess the skills and determination to succeed. Those who welcome challenges are more likely to cultivate perseverance and achieve a sense of accomplishment—each hurdle they overcome strengthens their capacity to manage future difficulties.

Embracing a challenge-oriented mindset can be particularly transformative for autistic adults. Many face unique challenges, including difficulties with social communication, sensory sensitivities, and adjusting to changes in routine. However, viewing these challenges as opportunities for growth rather than obstacles can significantly enhance their ability to navigate the world. By focusing on problem-solving and creative strategies, they can find ways to adapt, learn, and excel in various areas of life.

James, the 28-year-old graphic designer we discussed earlier, has struggled with social interactions. He often feels overwhelmed in workplace environments that require extensive collaboration and small talk. Early in his career, he avoided team projects, fearing that his difficulties with verbal communication would be seen as incompetence. However, instead of allowing these challenges to define him, James was able to reframe his perspective—he viewed his discomfort in social situations an opportunity to develop new skills, not as a roadblock. Determined to improve, James sought alternative ways to communicate that played to his strengths. He became proficient in written communications, using emails and project management software to convey his ideas clearly. He also began attending small, structured networking events to connect with others in a more controlled setting. Over time, his confidence grew, and he realized that his deep focus and attention to detail made him an invaluable team member. By embracing his challenges, James enhanced his communication skills and gained a greater sense of self-confidence and professional fulfillment.

His story highlights how shifting one's perspective on challenges can lead to remarkable growth. Instead of viewing obstacles as barriers, James used them as stepping stones to enhance his skills and build resilience. This approach empowered him to navigate the complexities of the professional world and ultimately proved that when challenges are embraced, they can become powerful catalysts for personal and professional success.

Personal Control

Personal control, the final component of stress hardiness, centers on individuals channeling their energy toward aspects of their lives that *can* be influenced while accepting and releasing what lies beyond their control. This principle is critical for reducing feelings of helplessness and cultivating a sense of autonomy and resilience. When individuals focus on areas they do have some control over—such as their thoughts, decisions, and behaviors—they gain a sense of empowerment, which is essential for effective stress management.

Personal control encourages individuals to shift their attention from unpredictable external circumstances like other people's actions or unforeseen events to internal factors within their grasp. While someone cannot control an unexpected change at work or at home, they often have more control over their response to the situation than they may initially realize—for example, they can choose to adapt, learn new skills, or seek support. This sense of ownership builds confidence and reinforces the belief that individuals can manage challenges effectively. Letting go of the need to control aspects of life that are beyond our control also reduces unnecessary mental and emotional strain. It liberates us from the burden of attempting to fix everything around us and instead enables us to focus our energy on constructive and impactful efforts. For autistic adults, developing a sense of personal control can be especially empowering. Many encounter unpredictability in social interactions, workplace dynamics, or changes in routine that can lead to increased anxiety, but they can build confidence and alleviate stress by focusing on what actions they *can* take, such as developing self-care routines and utilizing specific communication methods.

Emily, a 32-year-old woman with ASD, works as a data analyst. She thrived in structured environments but has struggled with sudden changes, particularly in the workplace—she found it overwhelming when meetings were rescheduled, priorities shifted, or managers expected immediate adaptability. For years, these unpredictable disruptions sent her into a spiral of anxiety and made her feel out of control. She often became frustrated, wishing she could better manage her environment to maintain stability.

However, with the assistance of a therapist, Emily eventually acknowledged that while she couldn't prevent last-minute changes at work, she *could* manage how she reacted to them. She started using strategies to cope with her stress, such as requesting written summaries of new assignments to better process changes. She also established a personal routine that provided her with a sense of structure, including scheduled breaks, using noise-canceling headphones when the office felt overwhelming, and practicing mindfulness techniques to restore her sense of calm. Instead of resisting or becoming overwhelmed by every unexpected change, Emily learned to anticipate disruptions and developed an action plan to navigate them. She became more confident in handling workplace stress by focusing on what she could control: her reactions, her workspace setup, and her communication strategies.

The principles of stress hardiness—commitment, challenge, and personal control—are especially pertinent for adults with ASD, who frequently face unique stressors in their daily lives. Social interactions, sensory sensitivities, and disruptions to routines can heighten stress, making it essential to develop customized strategies for resilience. When autistic adults align these three principles with their specific needs and strengths, they can cultivate stress hardiness, which will in turn serve as a powerful tool for fostering their well-being and adaptability. An individual with ASD who learns to embrace structured routines for self-care, reframe challenges as learning opportunities, and focus on managing their sensory environment may find themselves thriving in situations that had once seemed overwhelming.

The Stress of Social and Cognitive Challenges for Adults with ASD

Individuals with ASD often face heightened stress due to challenges in coping with complex social and cognitive processes. These difficulties extend beyond mere misunderstandings—they can lead to frustration, anxiety, exhaustion, and feeling overwhelmed by daily interactions and obligations. Grasping how these stressors impact autistic adults provides insight into the emotional burden of living in a world where social norms and expectations often seem unattainable.

The Stress of Understanding the Perspectives of Others

One of the most persistent stressors for individuals with ASD, as we noted in Chapter One, is their difficulties with "theory of mind." This theory represents the ability to understand that other people have thoughts, emotions, and intentions that are different from one's own. Misinterpreting others' emotions or intentions can lead to constant uncertainty, making social situations unpredictable and exhausting. This struggle often results in withdrawal, as autistic individuals may avoid interactions rather than risk further misinterpretations, and that in turn leads to feelings of loneliness and isolation.

Larry, a 26-year-old biology graduate student with ASD, felt comfortable while doing experiments in the lab—collecting data and discussing results with other students gave him a structured comfort zone. However, when the same students asked him to go out for a drink or join them to watch a game, Larry's anxiety grew. At times, he made excuses for why he couldn't accept their invitations. When he did go out with them, his mistimed humor or his attempts to steer discussions to their research left him feeling isolated, not able to engage in or understand non-research-related conversations. He had difficulty understanding why they didn't laugh at what he thought were his funny remarks. These events outside the lab even began to affect his relationships with the other graduate students at school.

The stress of navigating complex interactions can also impact self-esteem. Persistent social misinterpretations may cause individuals with ASD to feel inadequate, leading them to question their social competence. This ongoing struggle can contribute to heightened anxiety and make it even more challenging to engage with others. While strategies like structured social scripts and cognitive behavioral approaches can provide some relief, the need to maintain continuous vigilance in social settings remains a significant source of stress for autistic adults. But here's the good news: supportive environments that encourage direct and clear communication ease some of these difficulties. Colleagues, friends, and family members who practice patience and provide explicit feedback can make a meaningful difference. By fostering understanding and inclusivity, society can help alleviate the immense emotional toll that individuals with ASD face when trying to navigate social expectations.

The Stress of Emotional Connection

Engaging in joint attention of emotion—i.e., acknowledging and responding to another person's feelings—presents another major stressor for individuals with ASD. Even when they deeply care about others, expressing or interpreting emotions in a way that aligns with social expectations can be difficult. Miscommunications in these situations can lead to unintentional misunderstandings, strained relationships, and increased anxiety around social interactions. Many autistic individuals struggle to recognize unspoken emotional needs, making it challenging for them to provide the kind of support that others expect.

For example, Jill, a 30-year-old woman with ASD, struggled to comfort her roommate, Lisa, who expressed feelings of being overwhelmed about a recent work project. Wanting to be helpful, Jill attempted to offer logical solutions, listing ways Lisa could better manage her time and workload. However, Lisa wasn't looking for solutions—she wanted validation and emotional support. Because Jill's responses felt cold or detached, Lisa began to withdraw from their usual conversations. Jill, in turn, became anxious and confused, worrying about what she might have said that had apparently upset her friend. The stress of not knowing how to "fix" their interactions added to Jill's ongoing social anxiety and made every conversation feel fraught with potential pitfalls.

Over time, individuals like Jill may develop coping strategies, such as learning scripted responses to use in emotional situations or asking direct questions to clarify how others feel. However, the pressure to constantly analyze and adjust their reactions remains a significant source of stress.

Support from understanding colleagues, friends, and loved ones who communicate their needs more directly can help alleviate some of this burden, making it easier and less overwhelming to form emotional connections.

The Stress of Planning and Prioritization

Difficulties in executive functions like planning and attention to detail can lead to overwhelming stress, whether in the home or work environment. That's especially true when multitasking and flexibility are necessary. While many autistic adults may excel at focusing on specific details, they may struggle to manage interconnected or shifting responsibilities, increasing their feelings of frustration and burnout. The unpredictability of deadlines, changing expectations, or the need to transition between tasks can make routine responsibilities seem insurmountable.

Dan, a 27-year-old software engineer with ASD, typically hyper-focused on one aspect of a project while unintentionally neglecting others. When his team shifted priorities unexpectedly, he found it difficult to adapt—adjusting his workflow required significant mental effort. The unpredictability of these changes created a constant sense of stress, making him question his ability to succeed in collaborative environments.

With the assistance of a caring supervisor, Dan began to use structured task-management tools when having to shift from one activity to another. While these tools were helpful, his underlying anxiety about navigating an ever-changing workplace remained a challenge. Clear expectations and structured routines are crucial for adults with ASD (and arguably for all workers), yet many workplaces thrive on flexibility and rapid adjustments, making it difficult for individuals like Dan to feel comfortable.

Another example of the struggles that emerge when autistic individuals are faced with prioritizing commitments in their lives is illustrated by Madison, a 22-year-old college student who was struggling with organizing her coursework. She could dedicate hours to perfecting one assignment while completely forgetting about another one that was due the same day. Group projects were particularly stressful, as coordinating with others and managing multiple tasks simultaneously felt chaotic. When her professors changed deadlines or introduced new material unexpectedly, she experienced significant distress and wound up procrastinating or avoiding the assignment. Despite using detailed planners and reminders, the effort she had to expend to balance her priorities often left her mentally exhausted.

Dan and Madison put in much time and effort in an attempt to develop strategies to mitigate their struggles—they broke tasks into smaller steps, used visual schedules, and sought clarity from supervisors or professors—but the constant cognitive load of maintaining structure in an unpredictable world remained a source of stress for them. They and many others with ASD require supportive workplaces and academic environments that allow for structured schedules, advanced notice of changes, and explicit communications. When these factors are in place, they help reduce the anxiety that accompanies planning and prioritization challenges. By implementing small but significant adjustments, organizations and educators can create a more inclusive space for autistic individuals to thrive.

The Stress of Rapid Social Processing

Navigating unspoken social norms requires an intuitive processing speed that many individuals with ASD find bewildering. Analyzing and synthesizing social information can be mentally exhausting, particularly in environments that demand quick, seamless interactions; the pressure to perform in these settings can prompt feelings of inadequacy and heightened social anxiety. Many autistic individuals need additional time to process verbal and nonverbal cues, but the fast-paced nature of social exchanges often does not allow for this, leading to frequent misunderstandings or feelings of exclusion.

David, a 35-year-old marketing analyst, struggled in workplace meetings where discussions moved quickly—when the conversation shifted to a new subject, he had not yet fully processed or formulated a response to the prior topic. This recurring experience left him feeling frustrated, self-conscious, and disconnected from the team. Not surprisingly, over time, he became increasingly hesitant to contribute, knowing that he was out of sync with his colleagues.

David experienced the same issues at home with his wife, Rachel. Although she displayed understanding and patience, the recent birth of their first child had disrupted the established predictable flow of communication between David and Rachel. When their newborn son cried and needed his diaper changed or had to be breastfed, Rachel had to quickly attend to him. Although David loved being a father, he had difficulty being interrupted as he was talking with Rachel.

The unpredictability of his social interactions both at work and at home remained a significant source of anxiety for David. Rachel and David sought parent/couples counseling with a clinician knowledgeable about ASD. While the focus of the sessions was on improving their relationship as spouses and parents, David found that what he learned in counseling was also helpful at work: pausing discussions, summarizing key points, and creating structured turn taking helped him feel more confident in both social and professional settings. Fortunately, colleagues at work as well as Rachel at home fostered a more supportive and accommodating environment. And very importantly, David became more confident as a father.

Managing the Stress of ASD-Related Challenges

While these challenges create significant stress, many autistic individuals develop coping strategies to mitigate their impact. Techniques such as structured communications, explicit social expectations, and self-advocacy can help reduce feelings of unpredictability and anxiety. However, as we've already emphasized, having to constantly adapt to a world that does not always accommodate neurodivergent processing takes an emotional toll.

To foster a more supportive and inclusive society, all of us must understand the stress caused by these difficulties. By acknowledging the unique pressures faced by adults with ASD, individuals and management in workplaces, educational institutions, and social circles can implement accommodations that alleviate rather than exacerbate stress. Ultimately, the goal is not just for individuals with ASD to manage their stress but for the world around them to create environments free of negative judgment, surroundings where autistic adults—and everyone— can thrive without having undue emotional burdens placed upon them.

Finding Purpose through Commitment

One of the most transformative steps toward resilience is discovering a sense of purpose, also known as commitment, which is one of the three components of stress hardiness. For adults with ASD, this often involves identifying and leveraging personal strengths and interests to create meaning and fulfillment in daily life. Purpose not only provides motivation but also serves as a stabilizing force, offering direction and reducing the impact of stress.

Many autistic adults possess specialized interests or unique talents that bring them joy and satisfaction. Whether rooted in art, technology, science, writing, or other pursuits, these passions can be a robust foundation for building purpose and resilience. By channeling these interests into productive or creative activities, individuals can foster a sense of achievement, self worth, and emotional well-being.

Leila, a 33-year-old adult with ASD, discovered immense joy in creating intricate digital art. Initially hesitant to share her work, Leila kept her designs private, fearing judgment or misunderstanding. Over time, though, she mustered the courage to post her creations on social media, where she unexpectedly found a supportive community of artists who admired her distinctive style. With their encouragement, Leila began selling her artwork, transforming her passion into a source of income and personal fulfillment. This journey enhanced Leila's sense of belonging, significantly reduced her stress, and deepened her connection with others.

For other adults with ASD, their purpose may emerge from contributing to a meaningful cause, pursuing an area of study, or engaging in calming hobbies that provide inner peace. The key is to explore what feels genuinely meaningful on an individual level, free from the pressure to conform to societal expectations. This journey toward finding purpose can empower individuals to build resilience, leading to a life filled with intention, connection, and personal satisfaction.

Reframing Challenges as Opportunities

Life is filled with challenges, but how we perceive them significantly influences our ability to cope and thrive. For autistic adults, reframing difficulties as opportunities for growth is a powerful strategy to build resilience and foster a positive outlook on life.

This shift in perspective involves stepping away from the idea of failure and instead focusing on self-improvement and potential for personal growth. For example, a workplace misunderstanding that leads to conflict can be reframed as an opportunity to refine communication skills, learn to set boundaries, or advocate for accommodations that support success. Although reframing usually isn't easy—especially for autistic adults—reframing can be a very important step toward promoting well-being.

Tom, a 35-year-old adult with ASD, often felt frustrated and confused by the nuances of workplace social dynamics. This led him to change jobs several times. After a particularly challenging interaction with his current supervisor, however, he sought advice from a mentor. Together, they identified strategies to express his needs more clearly and navigate complex workplace relationships more effectively. With his mentor's support, Tom sat down with his supervisor to discuss some of his difficulties related to ASD and strategies that might help him be more successful at work. He was pleasantly surprised by the empathy his supervisor expressed. Over time, this proactive approach not only improved his professional interactions but also significantly reduced his stress and anxiety, enabling him to feel more at ease in his role.

Reframing challenges as opportunities doesn't mean minimizing the difficulty of a situation. Instead, it involves validating the stress while consciously shifting our focus to what *can* be learned or gained. This mindset fosters adaptability and encourages us to approach problems with curiosity and determination rather than fear or avoidance. It creates room for growth and resilience, transforming potential setbacks into opportunities for personal development.

Cultivating Personal Control

A sense of personal control is essential for managing stress effectively. For all adults and especially those with ASD, cultivating this control involves focusing on areas of life that can be influenced or adjusted, such as creating sensory-friendly environments, establishing routines, or advocating for personal needs.

Sensory sensitivities are a common source of stress for many autistic individuals, but proactively managing these triggers can significantly enhance daily well-being. Identifying specific sensory challenges and implementing targeted solutions is a key step. Emily, the data analyst we met earlier in this chapter, found her shared office space overwhelming due to constant noises and harsh fluorescent lighting. After recognizing the negative impacts on her concentration and emotional state, she requested noise-canceling headphones and a desk lamp with softer lighting. These relatively simple changes made a substantial difference and enabled her to work more comfortably and efficiently.

As Emily's situation illustrates, developing self-advocacy skills is another vital component of personal control. For autistic adults, learning to articulate their needs—whether at work, in relationships, or in public spaces—can help create more supportive and accommodating environments. We saw the growth of self-advocacy skills with Tom: while building these skills with the input of a mentor, Tom became more confident and persistent. Gaining the ability to set boundaries, request accommodations, and express preferences allows individuals to feel more empowered and gives them a greater sense of autonomy. All of this significantly reduces stress in their daily lives.

Additionally, cultivating personal control involves individuals focusing their energy on manageable aspects of life rather than dwelling on situations that feel overwhelming or uncontrollable. By making deliberate choices and seeking practical solutions, adults with ASD can create environments and routines that support their unique needs, paving the way for greater resilience and emotional well-being.

Emotional Regulation and Resilience

Emotional regulation is a cornerstone of resilience. That's especially true for adults with ASD, who often face heightened emotional responses to sensory or social stressors. Effectively managing these emotions is vital for navigating daily challenges and maintaining overall wellbeing.

Techniques such as deep breathing, progressive muscle relaxation, and mindfulness are powerful tools for emotional regulation. These strategies help calm the nervous system, lower physical tension, and create a sense of grounding during overwhelming moments. Mindfulness practices such as focusing on one's breath or engaging in a sensory grounding exercise can help individuals regain control when emotions feel intense. Over time, consistent use of these techniques makes it easier to handle stress and strengthens emotional resilience, enabling people to recover more quickly from challenging situations.

Equally important is the role of a strong support network. Trusted individuals—family members, friends, mentors, peers in an ASD-focused community—can provide empathy, encouragement, and practical advice during difficult times. Alex, whom we met earlier, was experiencing emotional meltdowns caused by sensory overload, but then he discovered the value of having a network through a local support group. Within this group, members shared personal experiences and coping strategies, offering Alex guidance and reassurance. This sense of belonging and understanding reminded him that he wasn't alone in his struggles, a realization that significantly enhanced his emotional resilience.

Developing self-awareness about emotional triggers can further improve emotional regulation. By recognizing patterns and preparing for potentially stressful situations, autistic adults can confront challenges with greater confidence and calmness. A combination of self-regulation techniques, supportive relationships, and self-awareness lays a strong foundation for emotional resilience, empowering individuals to navigate life's complexities more easily and with greater assurance.

Integrating Strengths and Addressing Challenges

Developing resilience requires addressing the unique challenges associated with ASD while simultaneously leveraging individual strengths. This process involves adopting practical strategies such as improving social understanding, fostering emotional connections, and establishing routines that create stability and reduce uncertainty. These strategies empower individuals to manage stress more effectively and simultaneously build a foundation for personal growth.

For adults with ASD, resilience is not about conforming to societal norms or suppressing their individuality. Instead, it's about embracing their authenticity, recognizing their unique contributions, and finding ways to thrive in environments that may not always accommodate their needs. Taking this approach includes celebrating personal strengths, whether in creative problem-solving, attention to detail, or specialized interests.

Becoming stress-hardy entails developing patience, practicing self-compassion, and adopting a mindset of continuous growth. Resiliency *is* attainable! It's also transformative when paired with access to the right tools, supportive networks, and tailored strategies. Resilience enables autistic adults to lead meaningful and fulfilling lives while confidently navigating challenges.

Five Key Takeaways

Stress Hardiness as a Mindset

Stress hardiness involves adopting a mindset of resilience rather than avoidance. It includes three core components: commitment, challenge, and personal control. By reframing stress as an opportunity for growth, individuals with ASD can better navigate difficulties, build confidence, and cultivate emotional well-being in their daily life.

The Power of Commitment

Commitment provides a sense of purpose and stability, helping people stay grounded during stressful periods. Finding meaning in work, relationships, or hobbies enhances resilience by offering motivation and direction. This connection to meaningful pursuits enables autistic individuals to manage stress effectively and focus on long-term personal fulfillment.

Reframing Challenges as Growth Opportunities

Viewing challenges as opportunities rather than threats allows individuals with ASD to develop adaptability. Instead of fearing setbacks, they can approach difficulties with problem-solving and curiosity. This shift in perspective reduces the tendency to catastrophize, enabling people to gain confidence and proactively manage stressors in various environments.

Personal Control Reduces anxiety

Focusing on what *can* be controlled—thoughts, routines, and reactions—empowers individuals to manage stress effectively. Accepting what can*not* be changed minimizes frustration and mental strain. By taking proactive steps like creating structured environments or practicing self-advocacy, autistic adults can navigate daily life with greater autonomy and confidence.

Emotional Regulation Strengthens Resilience

Managing emotions through mindfulness, deep breathing, and support networks enhances resilience. Recognizing emotional triggers and developing coping strategies helps individuals with ASD maintain balance in stressful situations. By fostering their self-awareness and using calming techniques, they can improve their emotional well-being and respond to stress with greater stability and self-assurance.

Self-Guided Activities to Choose the Path to Become Stress-Hardy

Activity 1: Building Commitment and Finding Purpose and Meaning

Identify what matters to you most and why, then actively channel your personal values into meaningful actions by listing your strengths and setting specific goals based on each strength, like using creativity to make art. When you integrate small habits like gratitude journaling and reflecting on your accomplishments into each day, you'll have a greater sense of commitment and intention. These practices will allow you to cultivate long-term motivations and a deeper connection to your identity and goals, especially during periods of doubt or stress.

Activity 2: Reframing Challenges as Opportunities

Reflect on recent stressful experiences you've had, then determine the lessons you've learned. This process will allow you to transform your frustrations into insights. By journaling daily, you can record your challenges, your responses to those challenges, and ways that you could potentially improve. This practice will strengthen your adaptive thought processes. You can also choose a specific recurring issue and brainstorm about it with the goal of arriving at three proactive solutions that will enhance your choices and their readiness. For example, addressing sensory overload could involve using noise-canceling headphones, scheduling tasks during quieter times, or establishing a calming post-activity routine. By reframing stressors as opportunities to learn, you'll build resilience and cultivate the belief that challenges can serve as pathways to personal growth, adaptability, and enhanced problem-solving capabilities.

Activity 3: Strengthening Personal Control

Creating a "control inventory" will help you separate actionable concerns (i.e., managing your personal schedule) from uncontrollable ones (i.e., the opinions of others). Once you've established your inventory, then you can design a routine that balances responsibilities, self-care, and enjoyable activities. The final step is advocacy: identify one need and draft an explicit, respectful request, such as asking for sensory breaks during meetings. These strategies will enhance your autonomy and confidence, helping you feel more grounded, organized, and assertive about navigating your daily challenges even as you're building greater control over your environment.

Activity 4: Cultivating Resilience through Emotional Regulation

Track your emotional responses throughout the day to identify your patterns and triggers, then explore calming strategies like mindfulness, grounding, or deep breathing. Select one to practice during stressful moments. Finally, spend five minutes visualizing yourself calmly navigating a challenging situation. This will reinforce your confidence and emotional control. These techniques give you tools to manage intense scenarios, reduce overwhelm, and promote more internal stability. By practicing them consistently, you'll develop a reliable emotional toolkit that will support your resilience in high-pressure or overstimulating environments.

Activity 5: Enhancing Support Networks

Begin by listing three trusted allies you can turn to for encouragement or advice, then explore local or online communities tailored to your needs or interests, like support groups for autistic adults. Finally, engage in some collaborative planning, such as working with a friend or family member to manage a recurring stressor (i.e., organize sensory-friendly grocery shopping trips). These activities promote connections, reduce isolation, and reinforce how important supportive relationships are for mental health, resilience, and personal growth.

Activity 6: Embracing Self-Compassion and Growth

Write a compassionate letter to yourself, responding to a recent challenge and offering encouragement and empathy, and track your small victories (i.e., navigating a social event or completing a challenging task) to celebrate your incremental progress. To reinforce a positive mindset, create a personal affirmation that highlights your strength and adaptability, such as saying aloud, "I can handle challenges with patience and determination." Such reflections promote self-acceptance and help combat self-criticism; you'll reinforce a mindset that values growth over perfection and progress over performance.

Activity 7: Reducing Overwhelm through Simplification

Practice breaking large tasks into smaller, concrete steps—for example, clean one room at a time instead of tackling the entire house. Focus on decluttering your physical space to minimize distractions and enhance concentration, and intentionally schedule downtime for hobbies, rest, or quiet reflection. These changes foster mental clarity and emotional relief; they'll reduce your feelings of overwhelm and promote an improved internal balance. Simplifying your responsibilities and environments will establish a foundation for better daily functioning and more sustainable well-being.

Activity 8: Reflecting on Progress

Each week, assess what went well and what needs adjustment, particularly with respect to your resilience and stress management. Evaluate the effectiveness of your strategies and brainstorm alternatives when something isn't working. Finally, choose a small, realistic task that will support your long-term development, like initiating a conversation to build your social confidence, and set that as a goal. Regular reflection will help you adapt to various situations more easily and will ensure that your efforts align with your evolving needs. In short, you'll sustain your momentum toward self-improvement! Over time, this process will foster insights, motivation, and a proactive mindset, all of which lead to continuous personal development.

Adopting these structured strategies will allow you to develop a path toward resilience, empowerment, and a stress-hardy mindset that's tailored to your strengths and needs.

4

Viewing Life through the Eyes of Others

Empathy is the ability to understand the world of others on both cognitive and emotional levels. Empathy allows us to connect with others, communicate effectively, and foster understanding in our relationships. Since individuals with ASD process social and emotional information differently from those without ASD, they struggle with developing comfortable interpersonal relationships.

The challenges that autistic adults face in their interactions with others do not arise from a lack of care or concern but rather from differences in neurological processing. These challenges can be linked to various cognitive and neurological factors, including variations in theory of mind, executive function, sensory processing, and social motivation. Fortunately, empathy is not an all-or-nothing trait; rather, it's a skill that can be nurtured and developed over time regardless of where one starts.

For example, James, the 28-year-old graphic designer with ASD discussed earlier, genuinely cared about his colleagues but often struggled to interpret their emotional states accurately. When his coworker Laura appeared distant during lunch, James assumed she simply wanted to be left alone, so he didn't say anything to her. Ian, another coworker, joined their table and noticed Laura's expression. He asked her with a caring tone of voice, "Is everything okay?"

Since Laura felt comfortable with both James and Ian, she shared that she had just learned that a friend was facing a serious health situation. Ian replied, "I'm so sorry to hear about your friend's illness! If there's anything I can do to help, please let me know."

James echoed Ian's comments, wondering why *he* hadn't recognized that Laura wasn't being as expressive as she usually was. As a result, he said nothing to her.

Difficulty with Theory of Mind

One of the main reasons that being empathic can be challenging for individuals with ASD is their difficulty with what we earlier called "theory of mind." ToM refers to the ability to attribute mental states—thoughts, beliefs, intentions—to oneself and others. Individuals with ASD often struggle with this ability, and that makes it difficult for them to assess how others are feeling or thinking. In the absence of an accurate ToM, interpreting social interactions becomes a significant challenge since autistic people may not appreciate that others have perspectives different from their own.

Simon is a 36-year-old accountant with ASD. When Alan, his supervisor, praised a report Simon had prepared and provided constructive feedback about a couple of details, Simon assumed that Alan was disappointed with all of his work. Given Simon's struggles to understand that feedback is meant to help rather than criticize, he became increasingly anxious and withdrawn. Alan quickly noticed this change in Simon's demeanor and asked if something was wrong.

Simon immediately said no, but Alan persisted—he knew he had a good relationship with Simon. Alan's persistence resulted in him explaining to Simon that similar to the feedback he was giving to all of the staff, his suggestions were meant to enhance an already solid report.

Having a supportive, empathic supervisor allowed Simon to gain a clearer understanding of Alan's perspective. If Alan hadn't understood and empathized with Simon's reaction, the latter's anxiety about his performance would have continued.

Impaired Joint Attention of Emotion

Another factor that contributes to the difficulty that autistic individuals have with empathy is impaired joint attention of emotion. This refers to the ability to share or acknowledge the emotional experiences of others. Joint attention assists individuals in recognizing and responding to the emotional states of those around them. For people with ASD, challenges in joint attention can lead to difficulties with recognizing nonverbal cues like facial expressions, tone of voice, or body language, all of which are essential "vocabulary" for understanding the feelings of others.

We saw this difficulty with joint attention of emotion when James failed to understand Laura's expressions. Another example involved Olivia, a 29-year-old teacher with ASD. During a conversation with her friend Emma, Olivia didn't notice that Emma's voice had become quieter and her facial expression had changed. Although the conversation didn't involve Emma's recent breakup with her boyfriend, Emma had suddenly thought about him. But she continued speaking with Olivia, even as she grew sadder. Olivia kept focusing on Emma's words rather than on her nonverbal cues. Later, when Emma explicitly told Olivia that she was upset,

Olivia felt guilty for not having recognized her friend's distress earlier. This difficulty in perceiving emotional signals is common among autistic adults and can make social interactions more challenging.

Executive Function and Empathy Challenges

Executive function deficits also play a role in challenges related to empathy. Executive function refers to cognitive skills that include planning, working memory, impulse control, and mental flexibility. These skills are essential for processing social information and adapting one's behavior accordingly. Individuals with ASD may struggle with shifting their focus from their own experiences to another person's perspective, making it more difficult for them to consider how someone else might feel in a given situation.

For instance, Justin, a 50-year-old engineer with ASD, has a structured way of approaching work and social interactions. When his colleague Tim was visibly stressed about an upcoming project deadline, given his own reliance on staying focused on a task, Justin assumed that telling Tim to do the same would be the best way to support him. However, what Tim actually needed was a validation of how he was feeling and an offer to help. But because of executive function challenges, Justin found it difficult to shift gears and consider an emotionally supportive response, and in responding to Tim, although his logical approach was well-intended, he came across as indifferent.

Eye Contact and Emotional Communication Struggles

Another important aspect of empathy is understanding the communicative content of a gaze. Eye contact and gaze direction convey a wealth of social information, including emotions and intentions. However, many autistic individuals find eye contact uncomfortable or overwhelming, which can make it hard for them to recognize subtle emotional cues. Without grasping the significance of gaze in social interactions, people with ASD may miss critical signals that guide appropriate emotional responses.

Rebecca is a 45-year-old artist with ASD who finds eye contact stressful and prefers to look at the ground or focus on an object while speaking. Her friend Samantha owned a small art gallery and was impressed with Rebecca's work, so impressed that she offered to display three of her paintings. Rebecca attended the opening day of the showing. While she wanted to speak comfortably with visitors, she had difficulty doing so and felt very anxious; she kept continually glancing down.

Samantha witnessed what was happening. Being aware of her friend's struggles, she caught Rebecca alone in a moment and suggested that rather than looking down while speaking with a visitor, Rebecca might look at the paintings. Since the discussions were about Rebecca's art, Samantha was hopeful that this strategy would lessen the impression that Rebecca wasn't interested in talking about her work. Rebecca followed Samantha's suggestion, and it did indeed improve her interactions with the attendees.

Sensory Processing and Empathy Challenges

Many autistic individuals experience sensory sensitivities that can make social interactions overwhelming—for example, loud noises, bright lights, or strong smells may create distress and distract from others' emotional cues. When an individual is coping with sensory overload, their ability to focus on and respond to another person's emotional state may be significantly diminished.

Alex, discussed previously, is a 45-year-old with ASD who experienced emotional meltdowns caused by sensory overload. During a conference, the loud chatter and bright lighting made it difficult for him to concentrate. When a couple of colleagues began to speak with Alex about some of the material being presented, Alex was too overwhelmed by the environment to fully engage with them. This inadvertently gave them the impression that he wasn't interested in what they had to say; he later realized that he had come across as indifferent even though he cared about his colleagues' remarks. Sensory challenges can significantly impact social interactions and empathy in such situations.

Problems with Emotional Regulation and Empathy

Finally, emotional regulation challenges can make expressing empathy difficult. Even when autistic individuals recognize and understand the emotions of others, they may struggle with managing their own emotional responses. If they're experiencing heightened anxiety or frustration, it can be particularly challenging to focus on responding empathically to the emotions of others. Fortunately, learning self-regulation techniques can be an essential step in developing more consistent and effective empathic responses.

Sean, a 34-year-old retail manager with ASD, experienced intense emotional reactions both at home and work when he felt stressed. When a coworker expressed her concerns about a difficult customer, Sean quickly felt overwhelmed, thinking of similar frustrating situations he had previously experienced. Rather than offering support or a possible way of dealing with this customer, he redirected the conversation to his own challenges.

Sean showed a similar pattern at home when his wife or children were facing problems— he would become very anxious himself, which only heightened the overall anxiety in the house. Eventually, with the help of a therapist, Sean learned to manage his emotions and respond in a more supportive manner.

Understanding these challenges is essential for autistic individuals to develop strategies that will support their ability to develop empathy. Fortunately, by recognizing the underlying cognitive and sensory differences between adults with ASD and those without, we can approach empathy development with patience and targeted interventions that cater to the unique needs of those on the autism spectrum.

Understanding the Perspectives of Others

The ability to understand another person's perspective is often taken for granted in social interactions. For many, it almost feels like second nature to sense how others feel and then adjust their own behavior accordingly. However, for individuals on the autism spectrum, this process can be far from intuitive. Andrew, a 20-year-old college student, was encouraged to join a robotics club on campus. During his first meeting, he approached a club member who walked away to the other side of the room before Andrew could say anything.

Given past situations in which he had been excluded, Andrew assumed that this other club member didn't want him there. His conclusion wasn't based on any evidence; rather, it stemmed from past rejections and a lack of clarity about the other person's emotions. In fact, this other member hadn't even noticed Andrew approaching—he had gone across the room to make an adjustment to one of the robots.

To develop a deeper understanding of how to grasp the perspectives of others, it's necessary to first acknowledge that everyone experiences the world differently. No two individuals interpret emotions, gestures, or words similarly—how someone expresses happiness, sadness, or frustration may differ based on their cultural background, personality, or past experiences. By recognizing this truth, autistic individuals can develop an appreciation for the inherent diversity in emotional expressions. This is a foundational step in building empathy.

The Role of Social Stories in Enhancing Empathy

Social stories are structured narratives that help individuals with ASD understand social interactions, emotions, and appropriate responses in different situations. These stories (developed by teacher and author Carol Gray) break down social experiences into clear, concrete, and relatable steps, giving individuals with ASD a framework for navigating complex interactions. By utilizing these narratives, autistic adults can better understand the emotions of others, recognize social cues, and develop appropriate empathetic responses.

Consider a social story that focuses on understanding a friend's disappointment. The story might begin with a scenario in which two friends, Taylor and Jordan, have planned to go out for dinner after work. Taylor is excited, but at the last minute, Jordan cancels due to feeling unwell. The social story describes how Taylor initially feels upset and might think, "Jordan doesn't want to spend time with me." However, the story then introduces an alternative perspective: Jordan is truly sick and is not avoiding Taylor. It encourages Taylor to respond empathically to Jordan by saying, "I'm sorry you're not feeling well. Let's reschedule when you're better." By reading and processing this scenario, autistic individuals can practice identifying emotions and responding with empathy.

Another effective social story might depict a scenario in which someone forgets a birthday. In this situation (again with Taylor and Jordan), Jordan expects a birthday message from Taylor, but Taylor forgets. Jordan feels hurt, but instead of assuming that Taylor doesn't care, the social story encourages Jordan to consider that people have busy schedules and make mistakes. The story then models an appropriate response, such as Jordan reaching out and saying, "Hey, I missed hearing from you on my birthday. I know life gets busy, but I'd love to celebrate with you sometime soon."

Social stories like these allow autistic adults to explore different perspectives, identify emotions in themselves and others, and practice forming responses that build connections rather than create conflicts. After reading these scenarios, role-playing them can further reinforce their effectiveness, helping individuals with ASD internalize empathic behaviors to use in real-life interactions.

Emotional Labeling and Expression

Emotional awareness is the foundation of empathy and effective social interaction. However, many individuals with ASD struggle to identify, label, and express their emotions, which can lead to frustration and misunderstandings. Being able to name emotions accurately allows us to process our feelings more effectively and communicate our needs to others.

One of the most effective ways to enhance emotional labeling is by using emotion charts. These charts visually categorize emotions, often ranging from basic feelings like happy, sad, and angry to more nuanced ones like overwhelmed, embarrassed, or content. For instance, an adult experiencing stress at work may initially describe their emotion as "angry," but an emotion chart might help them pinpoint a more specific feeling, like "overwhelmed" or "frustrated."

Recognizing this distinction enables them to express themselves more clearly and be able to say "I'm feeling overwhelmed with my workload and need a short break" instead of simply reacting with irritation.

Journaling is another powerful tool for enhancing emotional expression. Writing about daily experiences and associated emotions helps individuals identify patterns in their feelings and reactions. For instance, if someone regularly journals about feeling anxious before meetings, they may recognize that they struggle with public speaking anxiety rather than a general fear of work. This self-awareness allows them to seek support, such as practicing presentations in advance or requesting accommodations.

Moreover, role-playing exercises with trusted friends, therapists, or support groups can assist individuals in practicing how to express their emotions. For instance, a scenario might involve two colleagues discussing an office conflict. One person might say, "I feel left out when decisions are made without my input," while the other acknowledges their feelings and responds appropriately. This practice enhances emotional recognition and effective communication in real-world situations.

By regularly engaging in emotional labeling exercises and finding personalized methods of expression, autistic individuals can improve their ability to articulate their feelings, which leads to more meaningful and empathic social connections.

Nonverbal Communication and its Impact on Relationships

Nonverbal communication—including body language, facial expressions, tone of voice, and gestures—is vital to social interactions. Many individuals with ASD find it difficult to interpret these cues, which can make it challenging to understand the emotions and intentions of others. Developing the ability to recognize and respond to nonverbal signals can greatly enhance relationships within both personal and professional contexts.

Facial expressions are among the most important nonverbal cues. For example, a furrowed brow and crossed arms may indicate frustration, while a slight smile and a relaxed posture suggest friendliness. Autistic individuals can learn to recognize these cues by using flashcards featuring various expressions and identifying the emotions related to them. Watching television or movies without sound and trying to interpret the characters' emotions based on their body language can also be an effective exercise.

Tone of voice plays a significant role in communication as well. Sarcasm, for instance, can be difficult to detect, as it often involves a contradiction between spoken words and vocal tone. If a coworker says "Oh, great job!" with a flat or exaggerated tone after a mistake is made, an individual with ASD might take the statement literally. Practicing listening to recorded examples of different tones that are clearly identified—enthusiastic, sarcastic, neutral—can help distinguish these nuances.

Another valuable strategy is to engage in role-playing exercises, where one person expresses an emotion using only body language and the other attempts to interpret it. This activity builds awareness of nonverbal signals and improves responsiveness to subtle social cues.

By consistently practicing nonverbal communication skills, autistic adults can enhance their ability to understand unspoken messages, reducing social misunderstandings and allowing them to foster deeper, more empathic connections with others.

Practicing Empathy in Digital Communications

With the rise of text messaging, emails, and social media, much of modern communication lacks the face-to-face interactions that typically provide emotional context. This presents unique challenges for individuals with ASD, who may already be struggling with interpreting tones and implied meanings. Without nonverbal cues, messages can easily be misinterpreted and can lead to misunderstandings and unintended conflicts.

For example, a short text message such as "Sure" in response to a dinner invitation might be perceived as dismissive or reluctant, when in reality, the sender simply intended to confirm the plan. To navigate this, autistic individuals can benefit from using clarifying questions, such as responding, "Great! Looking forward to it. Does 7 PM work for you?" This encourages more expressive communication and helps ensure the intended meaning is clear.

Another effective strategy is incorporating emojis and punctuation to convey tone. A message that just says "Thanks" can seem cold, whereas "Thanks! " softens the tone and conveys friendliness. While some individuals with ASD may find emojis confusing or excessive, selectively using them can aid in expressing emotions and preventing misunderstandings. Email communications also present challenges, particularly in professional settings, where tone can be difficult to discern. Reading emails aloud before sending them can help ensure the tone is appropriate. If someone is uncertain whether their email may come across as too blunt, they can ask a trusted colleague or friend to review it prior to sending it.

Lastly, when participating in online discussions, it's beneficial to take a moment before replying to emotionally charged messages—pausing to consider the sender's possible intentions and crafting a thoughtful response can help avoid misinterpretations. For instance, if a friend shares a vague social media status that seems critical, instead of assuming it targets them, an individual might say, "Hey, I saw your post—just checking in to see if everything's okay." This method promotes empathy and encourages open communication.

By using these strategies, autistic individuals can learn to navigate digital communications more effectively, enhancing their online interactions and allowing them to build stronger personal and professional relationships.

Cultivating Empathy: A Path to Connections

Empathy is a vital skill that fosters meaningful connections and enhances social interactions. However, understanding and responding to the emotions of others can present unique challenges for autistic individuals. As we've explored in this chapter, difficulties with theory of mind, joint attention, executive function, and sensory processing can all contribute to differences in empathy development. Despite these challenges, empathy is not an unattainable or all-or-nothing trait—it's a skill that can be nurtured and refined through intentional practice and tailored strategies.

By examining real-world scenarios, we've observed the ways in which individuals with ASD can navigate social complexities and develop tools to enhance their empathy. From recognizing facial expressions and tone of voice to understanding nonverbal communication and managing digital interactions, various techniques can support an increased ability to build emotional awareness and understand that the perspectives of others differ from one's own. Social stories, emotional labeling exercises, and role-playing all provide structured ways for individuals to practice empathy in safe and guided environments. Additionally, self-regulation strategies assist in managing overwhelming emotions, allowing for more effective and compassionate interactions with others.

It's crucial to recognize that variations in how individuals express empathy don't indicate a lack of care or concern. Instead, those differences highlight the diverse ways that people process social and emotional information. By fostering an inclusive and understanding environment for everyone, we can support autistic individuals on their journey toward greater social confidence and connection.

Ultimately, viewing life through the eyes of others is an ongoing journey that requires patience, practice, and self-awareness. With the right tools and supportive networks, individuals with ASD can improve their ability to interpret emotions, engage meaningfully with others, and build strong, empathic relationships. By embracing these strategies, together, we can cultivate a world that values neurodiversity and fosters deeper understanding between all individuals.

Five Key Takeaways

Empathy Is a Skill that Can be Developed

Empathy is not an inherent, all-or-nothing trait but rather a skill that can be nurtured over time. While individuals with ASD may process emotions differently, they can improve their ability to understand and connect with others through practice and targeted strategies.

Challenges with Theory of Mind

Many individuals with ASD struggle with theory of mind (ToM), which is the ability to recognize that others have thoughts, feelings, and perspectives different from their own. This can lead to misunderstandings in social interactions but can be improved with guidance and structured learning.

Nonverbal Cues and Joint Attention Play a Crucial Role in Empathy

Difficulties in recognizing nonverbal signals—such as facial expressions, body language, and tone of voice—can make it harder for autistic individuals to interpret emotions accurately. Impaired joint attention of emotions further complicates social interactions and requires explicit communications for the sake of clarity.

Sensory Processing and Emotional Regulation Impact Social Engagement

Many individuals with ASD experience sensory sensitivities, which can make it difficult to focus on others' emotions during overwhelming situations. Emotional regulation challenges can also hinder empathic responses, making it essential to employ self-regulation strategies to improve social interactions.

Perspective-Taking Is the Key to Meaningful Connections

Understanding that emotions and behaviors vary based on individual experiences is critical for fostering empathy. By acknowledging that people express and interpret emotions in unique ways, autistic individuals can develop greater social awareness and build stronger, more meaningful relationships.

Self-Guided Activities to Cultivate Empathy and Perspective-Taking

Activity 1: Engaging in Perspective-Taking Reflection

Begin by noting your initial assumptions about a puzzling social interaction, then brainstorm at least two alternative reasons for the other person's actions. This technique promotes open mindedness and lessens quick judgments via encouraging you to reframe the situation and view it with empathy and curiosity. Engaging in this practice every week will help you develop the habit of looking beyond first impressions; in turn, you'll become more flexible with your social thinking. Over time, this kind of reflection will improve your communication skills and help you have more meaningful and considerate interpersonal relationships.

Activity 2: Doing an Empathy-Mapping Exercise

Enhance your emotional awareness by observing fictional characters and analyzing their nonverbal communication. Do this by pausing scenes in movies or shows to examine body language, facial expressions, and tone, then answer questions about the characters' emotions and how they're expressed. You can also imagine how you'd feel in a character's place. Once you're comfortable with fictional scenarios, you can transfer these skills to real-life interactions and continue to mentally note emotional cues. This will strengthen your empathy by sharpening your observational skills and the ability to consider the emotions and perspectives of others. With these tools, you'll be able to better able to understand how others feel in various social contexts and respond to them with greater sensitivity.

Activity 3: Cultivating an Awareness of Sensory Overload

Reflect on past situations where environmental stimuli like loud noises or bright lights interfered with your ability to connect with others, then identify specific sensory triggers and brainstorm strategies to manage them, such as using noise-canceling headphones or taking breaks. You can increase your self-awareness and comfort by selecting one strategy to implement in an upcoming social situation and reflecting on how well it worked (or didn't) after the fact. Effectively managing your sensory inputs will allow for clearer, more compassionate interactions with others and will promote having more balanced social experiences.

Activity 4: Practicing Theory of Mind: "What Are They Thinking?"

By watching a silent video or muting a conversation, you can focus on facial expressions and gestures to guess what the characters might be thinking or feeling. Then replay the scene with the sound on to check your interpretations. This will build up your skill with reading nonverbal cues and will make you more curious about the internal states of others. With practice, you can apply this technique during real-life conversations and ask yourself what others might be thinking. This will give you more attuned, insightful social interactions and a better emotional understanding of others.

Activity 5: Having Role-Playing Conversations

Role-play scenarios like supporting a stressed colleague and then switch roles to experience different perspectives. Afterward, discuss what was challenging and how your perceptions of the scenario changed. Such dialogues will uncover your assumptions and promote a better understanding of others' viewpoints. Practicing different social scripts each week will allow you to improve your confidence, listening skills, and compassion—through role reversal and reflection, you can gain deeper insights into interpersonal dynamics. This will allow you to handle real-life conversations with more empathy, patience, and effectiveness.

Activity 6: Writing in an Emotional Regulation Journal

At the end of the day, describe a social interaction that involved strong emotions, then examine your own feelings, the emotions of others, and any contributing stresses or sensory issues. Consider how you might respond to the same scenario differently in the future. This process will reveal emotional patterns and give you a greater awareness of your triggers and reactions. By tracking your responses over time, you can develop strategies for managing your emotions more easily and responding more empathetically. Your journal will become a personal growth tool that supports your emotional intelligence and enhances your social harmony.

By practicing these exercises regularly, you can enhance your ability to interpret emotions, connect with others, and develop stronger, more meaningful relationships.

5
Effective Communication: Tools for Navigating the Social World

Communication serves as a bridge between understanding and misunderstanding. For autistic individuals, this bridge often feels fragile, spanning the turbulent waters of unspoken cues, complex social norms, and subtle emotional undertones. Unlike neurotypical individuals who may instinctively grasp the nuances of social interaction, those with ASD often find these subtleties elusive, which leads to feelings of frustration and alienation. A simple conversation can become a daunting challenge as there's a risk of misinterpretation at every turn. Whether struggling to read facial expressions or decipher the hidden meanings behind specific phrases— or whether someone is feeling overwhelmed by the rapid pace of dialogue—communication barriers can affect relationships, work environments, and people's overall well-being.

In this chapter, we'll focus on providing individuals with ASD with tools to improve their communication skills. By examining strategies such as active listening, clear expression, and adjusting communication styles for different social contexts, we aim to empower readers to navigate their social environments with increased confidence and resilience.

Communication involves more than words—it encompasses body language, tone of voice, timing, and even silence. Depending on how it's delivered, a single phrase can inspire a multitude of interpretations. For someone with ASD, decoding these layers can be particularly challenging. Imagine standing in front of a puzzle where the pieces don't seem to fit together! A conversation can feel just like that. For instance, when someone says, "I'll think about it," a neurotypical person may interpret this statement as a polite way to decline an offer. However, an autistic person might take it literally and expect to receive a response later. These differences in interpretation can lead to confusion and frustration. Fortunately, autistic adults can use various strategies to piece together fragments of communication and create meaningful connections; with these tactics, they can ensure that communication is a tool for connection, not a barrier to it.

Why Communication Is So Challenging for People with ASD

Communication is a fundamental aspect of human interactions—it enables individuals to share thoughts, emotions, and information, thus fostering relationships and social connections.

Effective communication is crucial in various areas of life: personal relationships, the workplace, social settings, etc. However, for individuals with ASD, communication presents a unique set of challenges that affect both verbal and nonverbal exchanges. These difficulties arise from cognitive differences, challenges in social processing, and sensory sensitivities that make it more difficult to engage in everyday interactions.

As highlighted in the previous chapter about empathy, autistic adults may struggle with understanding social cues. They may also find it difficult to maintain fluid conversations and interpret nonverbal signals such as facial expressions and body language. Engaging in small talk, recognizing sarcasm, or understanding implied meanings in conversations can be challenging. As a result, they may experience misunderstandings, social anxiety, and difficulty forming and maintaining personal and professional relationships. Many individuals with ASD also experience sensory sensitivities that can make environments like crowded offices or social gatherings overwhelming, further complicating communications.

James, the 28-year-old graphic designer with ASD we mentioned earlier, is highly skilled in his field, but he finds workplace interactions challenging. He often misses subtle cues during meetings, such as when colleagues exchange glances that signal a need to change topics. He takes feedback literally and struggles with vague or indirect instructions, which results in misunderstandings with his manager. Additionally, James often avoids office social events because the noise and unpredictable conversations cause him anxiety. This affects his ability to connect with his colleagues and limits his professional networking opportunities.

To understand why communication is so challenging for individuals with ASD, it's essential to examine the specific areas where they experience difficulties and how these challenges manifest in their daily lives. By raising awareness of these issues and implementing accommodations such as clear communications, written instructions, and structured social interactions, workplaces and communities can better support autistic adults in navigating their social and professional environments.

The Ability to Attribute Mental States to Oneself and Others

One of the core challenges faced by individuals with ASD is the ability to attribute mental states—beliefs, desires, intentions—to themselves and others. This cognitive skill, which we described as "theory of mind", is crucial for effective communication because it allows individuals to predict and interpret social behaviors. Without a well-developed ToM, a person may struggle to understand why someone is upset, why a joke is funny, or how another person's perspective might differ from their own.

ToM allows individuals to recognize that others have their own unique thoughts and emotions, which may not always be explicitly stated. It plays a vital role in social interactions, from casual conversations to complex relationships. For neurotypical individuals, understanding social nuances comes naturally, as they can infer emotions and intentions based on subtle cues such as tone of voice, facial expressions, and body language. However, for autistic individuals, these inferences are not always intuitive. This difficulty can make interactions feel confusing, an outcome that leads to misunderstandings, social friction, and even unintentional offense.

For example, if a coworker sighs and says, "I can't believe how much work I have to do today," a neurotypical person may see that as an indirect request for empathy or help. They might reply, "That sounds really stressful. Do you need any assistance?" In contrast, someone with ASD may interpret the statement literally and view it as a simple observation rather than an expression of frustration or a plea for support. As a result, they might respond in a way that seems unempathic or unrelated, such as saying "Maybe you should manage your time better" or simply going back to their task without acknowledging their coworker's distress. This gap in understanding can complicate social interactions and potentially lead to strained workplace relationships.

The challenges associated with ToM extend beyond workplace interactions. In personal relationships, autistic individuals may struggle to recognize when a friend is feeling hurt or when a romantic partner expects emotional validation. For instance, if a friend cancels plans due to feeling exhausted, a person with ASD might assume that the cancellation is purely logistical rather than considering that their friend might need support or reassurance. Similarly, in romantic relationships, difficulties with understanding a partner's emotional needs can create frustration on both sides, as the neurotypical partner may feel ignored or misunderstood.

Additionally, ToM deficits can lead to difficulties in conflict resolution. Because individuals with ASD may struggle to see an issue from another person's perspective, it can be hard to navigate disagreements effectively—without grasping the underlying emotions involved, they might rely solely on logic and factual reasoning, and that can sometimes seem cold or dismissive to others.

Recognizing the impact of ToM deficits is essential for improving communication and social integration for autistic individuals. Structured interventions like social skills training, cognitive behavioral therapy, and explicitly learning perspective-taking strategies can help bridge gaps in communication. Additionally, neurotypical individuals can support their ASD peers by practicing clear, direct communication, avoiding ambiguous statements, and being patient with differences in social understanding. With increased awareness and accommodations, individuals with ASD can develop stronger interpersonal relationships and feel more included in their social and professional communities.

The Ability to Display Emotional Reactions Appropriate to another Person's Mental State

Joint attention of emotion is the ability to respond to another person's emotional state appropriately, and it's another area where individuals with ASD often struggle. Emotional reciprocity plays a significant role in communication—it helps people form bonds and navigate social interactions smoothly. However, autistic individuals may not instinctively mirror emotions or react in expected ways, which can lead to misunderstandings and strained relationships.

Emotional reciprocity involves recognizing, processing, and responding to emotions in real time, often through nonverbal cues such as facial expressions, tone of voice, and body language. While neurotypical individuals naturally engage in this process, individuals with ASD may have difficulty decoding these signals. As a result, they might either fail to react or they may display an emotion that seems out of place. This misalignment can lead others to assume that the autistic individual is uninterested, apathetic, or emotionally detached even when that's not the case.

For example, if a friend shares exciting news about a job promotion, a neurotypical person may respond with enthusiasm—they might reflect their friend's excitement by smiling, using an animated tone, or offering congratulatory words. In contrast, a person with ASD might reply with a flat tone, limited facial expressions, or a factual statement like "That sounds nice," which can seem unenthusiastic or dismissive. Likewise, when someone discusses a personal hardship, an individual with ASD may find it challenging to provide the expected words of comfort or might respond in a logical, problem-solving way instead of offering emotional support. Again, this can lead to social friction.

Conversely, some autistic individuals may exhibit emotional responses that seem exaggerated or misaligned with the context. For example, they might laugh at inappropriate moments due to difficulties in interpreting emotional cues or react with intense distress to minor frustrations. These variations in emotional expression can make it challenging to establish rapport and maintain relationships, as neurotypical individuals may misinterpret these responses as being a lack of empathy or social awareness.

This challenge highlights the importance of structured support (social skills training, emotional coaching) to help individuals with ASD recognize and respond to emotions in socially expected ways. Additionally, neurotypical individuals can foster better communication by being patient, using explicit verbal expressions of emotion, and understanding that differences in emotional responses do not necessarily reflect a lack of care or concern. With awareness and accommodations, people can build meaningful connections despite these differences in emotional reciprocity.

The Ability to Plan and Attend to Relevant Details in the Environment

Executive function deficits are common in individuals with ASD and can impact communication in various ways. Planning conversations, staying on topic, and filtering relevant details from irrelevant ones require a level of cognitive organization that many autistic people struggle with. These challenges can affect both verbal and written communication, making it more difficult to engage in smooth conversations, understand social expectations, and maintain coherence in discussions.

One key aspect of executive function is the ability to prioritize information effectively. Neurotypical individuals usually assess a situation and determine which details are most relevant to convey in a given context. Individuals with ASD, however, may struggle with this filtering process, leading them to focus excessively on minor details or present information in a way that seems disorganized or tangential. This difficulty can make conversations feel disjointed, as they may dwell on specifics that others find unimportant or fail to provide necessary background information.

Consider an autistic individual who's trying to engage in small talk the way Susan was trying to do. (She's the software developer we met in Chapter Two.) This person may struggle to recognize which details are relevant to share or may hyper-focus on a particular aspect of the conversation that others find insignificant. For instance, discussing weekend plans might lead to talking about the specific details of a bus schedule in great depth—listing particular times, routes, and transit delays, for example—rather than talking about the overall plans. While the information about bus schedules may be highly relevant to the individual with ASD, their conversation partner might find the conversation difficult to follow or disengage from. Additionally, autistic individuals may struggle to smoothly shift between topics. They may continue discussing a subject long after their conversation partner has lost interest, or they may abruptly change topics without providing sufficient context. This can make conversations feel one-sided and may lead neurotypical individuals to perceive them as being uninterested in reciprocal dialogue.

These challenges extend to professional and academic settings as well. In a workplace meeting, an employee with ASD may provide excessive detail when answering a question, thus inadvertently veering off-topic. Similarly, in an academic setting, they may struggle to summarize key points in an assignment and focus on peripheral details instead.

Supporting autistic individuals to help them improve their executive function skills can enhance their communication abilities. Strategies such as using conversation scripts, practicing summarization techniques, and receiving direct feedback on communication styles can help them develop a more transparent and more organized approach to social interactions. Meanwhile, neurotypical individuals can foster better communication habits by being patient, providing guidance on conversational flow, and recognizing that differences in information processing do not indicate a lack of engagement or intelligence. With a greater mutual understanding, conversations can become more balanced and fulfilling for both parties.

The Ability to Understand the Communicative Content of Gaze

Eye contact and gaze are potent tools in nonverbal communication—they convey emotions, intentions, and social engagement, and they play a crucial role in interactions. Neurotypical individuals rely on eye contact to signal interest, regulate conversation, and express empathy. However, many individuals with ASD find it difficult to interpret or sustain eye contact, which can lead to misunderstandings and social difficulties.

Due to sensory sensitivities, maintaining eye contact can be overwhelming or even physically uncomfortable for some autistic individuals. (We witnessed this being the case for Rebecca, whose paintings were being displayed at a friend's gallery.) Others with ASD may not naturally recognize the social significance of gaze and might meet the gaze of others infrequently or not at all during conversations. Neurotypical individuals might misinterpret this as disinterest, dishonesty, or lack of engagement even though the person with ASD may be fully attentive in their own way.

In addition to discomfort with eye contact, individuals with ASD may struggle to understand the communicative meaning behind gaze shifts. Neurotypical individuals often use eye contact during conversations to indicate that it's time for the other person to speak (when they briefly make eye contact) or to suggest that a conversation is ending (when they shift their gaze). Without the ability to interpret these cues, autistic individuals might interrupt frequently, speak for extended periods without realizing it's someone else's turn, or struggle to initiate conversations at appropriate moments. This difficulty can cause frustration for both parties and reinforce feelings of social disconnection.

Susan, the software developer, had difficulty when attending a work meeting. Although she was knowledgeable about the topics being discussed, she struggled to follow the subtle social dynamics of when to contribute—because she didn't recognize that her supervisor had shifted her gaze to indicate an open floor for discussion, Susan remained silent. Conversely, she sometimes began speaking at an inappropriate moment, not realizing that her colleague was preparing to make a point. These misalignments in social timing can make workplace interactions challenging and sometimes lead to misunderstandings about an individual's competence or engagement level.

Because eye contact plays such an essential role in communication, it can be beneficial for individuals with ASD to employ alternative strategies. For instance, they might find it helpful to look at a person's forehead or nose instead of directly into their eyes—that tactic can reduce their discomfort while still making them appear to be engaged. For their part, neurotypical individuals can improve communication by recognizing that a lack of eye contact does not necessarily signal disinterest or rudeness. By fostering environments that respect alternative forms of engagement (e.g., verbal affirmations or written communications), autistic individuals can navigate their social and professional interactions more effectively.

The Ability to Work Cooperatively with Others (Joint Attention of Behavior)

Collaboration and cooperation often require implicit social understanding and the ability to read between the lines. Joint attention of behavior—where two individuals focus on the same object or task through the lens of a shared experience—is an area where adults with ASD commonly struggle. This skill is essential in professional and social settings as it allows people to work seamlessly in teams, follow subtle cues, and engage in cooperative problem-solving. In contrast, when joint attention is impaired, it can create barriers to effective teamwork, cause frustration among colleagues, and lead to misunderstandings in social interactions.

In a professional environment, joint attention is crucial during everything from meetings and collaborative projects to daily workplace interactions. For instance, in a business meeting, team members often use subtle nonverbal cues like nodding toward a document or making eye contact with a specific colleague to indicate whose turn it is to contribute or what task should be prioritized. An autistic individual may miss these cues and instead continue focusing on unrelated aspects of the discussion, leading to workflow misalignment and causing team confusion. Similarly, in an open office environment, coworkers may use glances or gestures to signal the need for attention or input. Without the ability to interpret these signals, a person with ASD may unintentionally ignore requests or interrupt conversations at inappropriate moments. Social settings heavily depend on joint attention. For instance, individuals often rely on unspoken social rules to navigate interactions at a networking event. If a group subtly angles their bodies away from a conversation partner, it may signal that they are concluding the discussion. A neurotypical person would recognize this as a cue to step back, while an individual with ASD might not notice it and would perhaps continue speaking, accidentally creating an awkward situation. Similarly, if someone were to raise a glass at a dinner gathering, most people would interpret that as an invitation to participate in a toast. An autistic individual may not immediately recognize the gesture and might hesitate or not engage, causing others to perceive them as being socially aloof.

The challenges associated with a lack of joint attention can affect professional advancement and personal relationships—misinterpreting social dynamics in the workplace may lead to difficulties with forming professional connections, securing promotions, or working effectively within a team. In social situations, struggles with joint attention can result in people feeling excluded or finding it difficult to maintain friendships.

To help autistic adults enhance their joint attention skills, it can be beneficial to provide structured workplace accommodations, direct verbal communications, and explicit social coaching. Employers and colleagues can promote inclusivity by providing clear, direct instructions instead of relying on nonverbal cues, and they can also utilize written communication whenever possible and allow for organized turn-taking in discussions. Likewise, social groups can be more accommodating by recognizing that missed cues do not imply disinterest or rudeness, but rather reflect a different way of processing social interactions. With heightened awareness and minor adjustments, adults with ASD can navigate professional and social environments more effectively, enhancing their ability to collaborate and engage meaningfully with others.

The Ability to Understand, Comprehend, Analyze, Synthesize, Evaluate, and Differentiate Social Information

Social interactions are filled with nuances, implied meanings, and ever-changing contexts that require quick comprehension and adaptation. Neurotypical individuals usually process these complexities instinctively, using contextual cues, tone of voice, and body language to navigate conversations. However, for autistic adults, these social subtleties can be challenging to interpret, making everyday interactions more difficult and sometimes stressful.

One of the key challenges that individuals with ASD face is understanding nonliteral language, i.e., sarcasm, metaphor, and indirect communications. Many social exchanges rely on these forms of expression to convey humor, criticism, or emotional subtext. For instance, if a colleague says "That was a brilliant idea!" with a sarcastic tone after having made a mistake, most neurotypical individuals would recognize the sarcasm based on vocal inflection and context. However, someone with ASD may take the statement literally and miss the intended criticism or humor. This can lead to them feeling confused, misinterpreting social feedback, and struggling to respond appropriately.

Another common challenge is synthesizing and evaluating social information in real time. Autistic individuals may struggle to identify the unspoken rules of conversation, such as when to offer a reciprocal question, shift topics, or end a conversation. In professional settings, this can manifest as someone unintentionally monopolizing a discussion or failing to recognize disinterest. They may also speak too directly in situations that require diplomacy. For example, when asked for feedback on a colleague's presentation, a neurotypical individual might soften their critique with phrases like "I think you had some great points, but maybe you could clarify this part a little more," while an individual with ASD may provide a blunt response like "That wasn't clear at all." The latter response could unintentionally cause tension.

Furthermore, individuals with ASD may struggle to evaluate social hierarchies and norms. Workplace dynamics often require understanding implicit power structures—within these structures, how one communicates with a manager differs from one interacts with peers. Someone with ASD might not recognize the need for professional distance and might speak too informally with a superior, or they may struggle to gauge when a casual conversation transitions into a serious discussion. For example, during a team meeting, a neurotypical employee might detect a shift in tone when a manager begins discussing company goals and would adjust their demeanor accordingly. An autistic individual might miss the change in expectations and instead continue making informal or unrelated comments.

These difficulties can make social interactions feel unpredictable and overwhelming, which leads to anxiety and a reluctance to engage in workplace or social settings. Strategies such as receiving explicit social guidelines, practicing role-playing scenarios, and using direct and clear communication can help autistic adults navigate these complexities. Additionally, colleagues and supervisors can foster inclusivity by offering constructive feedback in a direct but considerate manner, avoiding ambiguous language, and recognizing that differences in social processing do not equate to a lack of intelligence or capability. With greater awareness and structured support, individuals with ASD can improve their ability to interpret social information and ultimately enjoy more effective communication and stronger professional and personal relationships.

The Foundations of Effective Communication

At its core, communication is about sharing ideas, feelings, and intentions.

Communicating is an interplay between expression and reception, a process where the speaker and the listener work together to convey and interpret meanings. Effective communication is not just about speaking clearly, it's also about listening actively, reading nonverbal cues, and responding appropriately. For individuals with ASD, however, each one of these aspects may require intentional practice and adaptation.

Consider Marsha, a woman we met in Chapter One who struggled to make friends because she often missed social cues. She would sometimes respond to a greeting with an unrelated comment or fail to notice when someone wanted to end a conversation. Over time, these missteps created barriers and left Marsha feeling isolated. Fortunately, with the assistance of a therapist, she discovered techniques to improve her communication skills through structured learning and practice, such as mirroring conversational partners' tones and recognizing when to pause and listen.

Active Listening: Building Bridges of Understanding

Active listening is fundamental to effective communication. It involves fully concentrating on the speaker, grasping their message, and responding thoughtfully. This skill transcends merely hearing words—it entails interpreting tone, acknowledging emotions, and giving appropriate feedback. Autistic individuals may need to make a concerted effort to participate in active listening since sensory sensitivities, challenges with nonverbal cues, or an intense focus on one's own thoughts can make it difficult to remain engaged in a conversation. One common challenge that individuals with ASD face when it comes to active listening is filtering out background noises and distractions. In busy environments like workplaces, restaurants, and social gatherings, competing sounds can hinder their ability to concentrate on the speaker. Additionally, some autistic individuals may focus so intently on specific details of the conversation that they overlook the overall message—they don't see the forest for the trees, as the saying goes. Instead, they may fixate on a particular word choice or factual detail instead of processing the speaker's emotional tone or intent.

Individuals with ASD can practice active listening by employing strategies that encourage engagement and understanding. Maintaining eye contact can be beneficial, but if direct eye contact feels uncomfortable, a neutral focal point can serve as an alternative—the speaker's forehead, a nearby object, etc. Another technique is to employ mindful listening, where the listener intentionally refocuses their attention on the speaker if they notice their thoughts drifting away.

Other tactics include responding verbally in the affirmative or summarizing the speaker's words—those strategies can reinforce understanding and demonstrate engagement. For instance, if someone says, "I'm feeling overwhelmed at work," an active listening response might be, "It sounds like you have a lot on your plate. Would you like to talk about what's been the most stressful?" This approach validates the speaker's feelings and encourages a deeper discussion, fostering feelings of connection and trust.

Active listening is essential for team collaboration, conflict resolution, and building relationships in professional environments. An autistic person who has difficulty with implicit social rules may unintentionally interrupt someone or misinterpret a speaker's tone, resulting in misunderstandings. Employing active listening techniques—such as pausing for a few seconds before responding, asking clarifying questions, and utilizing structured communication tools— can help bridge this gap.

Furthermore, neurotypical individuals can support those with ASD by being patient, using clear language, and allowing for extra processing time. Encouraging structured conversations where topics are explicitly stated can also make interactions more accessible. By fostering an environment where active listening is a mutual effort, both neurodivergent and neurotypical individuals can communicate more effectively, thereby strengthening relationships in personal as well as professional settings.

Clear Expression: Speaking with Purpose and Precision

Expressing thoughts and feelings can be incredibly challenging for individuals with ASD, especially when emotions are involved. Words may not come quickly, and others may misunderstand them due to differences in tone, facial expressions, or phrasing. Effective communication emphasizes clarity, simplicity, and intention, ensuring that messages are received as they are meant to be.

Autistic adults should begin by identifying the main point they wish to convey before speaking. Breaking complex ideas into smaller, manageable parts makes it easier for both the speaker and the listener to process the information, and using concrete language rather than abstract or figurative expressions can help prevent misunderstandings. It's also beneficial to keep in mind that when discussing emotions or personal experiences, using "I" statements can help reduce defensiveness and encourage constructive dialogue. For instance, instead of saying, "You're always ignoring me," a more transparent and less confrontational approach would be, "I feel left out when we don't talk as much." This change in wording fosters open conversation and minimizes conflict.

Clarity is also essential in professional environments. Being direct and specific about needs, boundaries, and expectations helps ensure that colleagues and supervisors understand the message. As an example, instead of saying, "I don't like how this project is going," a more precise statement would be, "To avoid confusion, I believe we should clarify our priorities before moving forward." If it's difficult to express thoughts verbally, individuals with ASD should write their thoughts down beforehand—then they'll better equipped to express themselves. Practicing these skills in low-stakes settings like during casual conversations with a trusted friend or family member can build confidence for more high-pressure interactions, such as workplace discussions or social gatherings. Over time, developing clear and purposeful speech can enhance relationships, improve professional interactions, and foster greater self-expression.

Adapting Communication Styles to Social Contexts

Every social setting has unwritten rules and norms that vary depending on relationships, professional hierarchies, and cultural expectations. Navigating these differences can be daunting, especially for autistic individuals, who may find it challenging to comprehend implicit social cues. However, adapting communication styles to fit different contexts is a valuable skill that minimizes misunderstandings and strengthens connections. For example, conversing with a close friend often entails casual language, humor, and shared experiences, whereas addressing a supervisor mandates a more professional tone, structured responses, and respect for workplace etiquette. Recognizing these distinctions is essential for effective communication.

One way to adapt to different social contexts is through observation. Individuals with ASD are advised to pay attention to how others interact in specific situations—they should notice their tone of voice, word choices, and level of formality and then practice mirroring these behaviors in similar settings while staying authentic in their communication. If they're unsure about the expectations in a particular environment, they should not hesitate to ask for clarification. For example, in a professional meeting, an individual with ASD might say, "Is this a good time to share my thoughts?" or "Would you prefer a written summary of my ideas?" This ensures their contributions align with the group's communication norms.

Role-playing scenarios can also be an effective way to practice adapting communication styles. Autistic people might work with a trusted friend, therapist, or support group to simulate different social situations, such as a job interview, a casual conversation, or a conflict resolution discussion. These exercises allow them to experiment with different tones, word choices, and levels of formality in a safe and supportive environment. Over time, with practice and awareness, adjusting communication styles to fit various social contexts can become a more intuitive process, leading to more confident and productive interactions in both personal and professional settings.

The Power of Nonverbal Communication

Nonverbal communication—body language, facial expressions, gestures, tone of voice— plays a significant role in conveying meaning, often complementing or even surpassing spoken words. While neurotypical individuals naturally integrate verbal and nonverbal cues, people with ASD may find it challenging to interpret or use them effectively. This can lead to misunderstandings, as others might misinterpret a lack of facial expression or unusual gestures as disinterest or disengagement.

That said, with practice and awareness, autistic adults can cultivate their nonverbal communication skills to enhance their social interactions and professional relationships. They might focus on their body language to ensure it aligns with their intended message, for example. When supporting a friend, maintaining an open posture, using a gentle tone, and occasionally nodding can help strengthen your ability to express empathy. On the other hand, maintaining a flat tone, avoiding eye contact, or crossing arms may unintentionally signal detachment or discomfort. Being aware of these cues can help adults with ASD ensure that their words and body language convey their true intentions.

Another strategy is to seek feedback from trusted individuals—a friend, mentor, or therapist can help identify specific nonverbal behaviors that need adjustment. They might point out if someone's tone of voice doesn't match their emotions or if their gestures seem to be out of sync with the conversation. It can also be helpful for people to record videos of themselves in conversation and then assess them.

Developing nonverbal communication skills takes time, but gradual practice can make interactions more natural. As autistic adults increase their awareness of nonverbal communication elements, these cues become easier to recognize and incorporate into daily interactions. This ultimately helps individuals build stronger social and professional connections.

Navigating Conflicts: Communication as a Tool for Resolution

Conflicts are an inevitable part of life, but they don't have to be destructive. Effective communication can transform conflicts into opportunities for growth, deeper understanding, and stronger relationships. For autistic individuals, managing conflict often necessitates putting forth an extra effort to regulate their emotions, accurately interpret social cues, and stay focused on reaching a resolution. Several aspects can complicate conflict resolution, ranging from differences in communication styles and challenges in recognizing implied meanings to difficulties with reading emotional subtexts. Nevertheless, with structured approaches, individuals can navigate conflicts in a way that encourages mutual understanding.

One of the initial steps in conflict resolution is to clearly identify the issue and express feelings using "I" statements. This approach minimizes defensiveness and promotes a productive conversation. Rather than assigning blame, directly stating emotions and needs can help avoid misunderstandings. For instance, instead of saying, "You never include me in plans," individuals with ASD might say, "I feel left out when I don't get invited to events. I would like to be included more often." This approach allows for a discussion without making the other person feel like they're being attacked.

Listening actively is equally important. Even if an autistic adult disagrees with the other person's perspective, they should resist the urge to interrupt or immediately counter their statements. Instead, it's better to focus on understanding the other person's point of view by asking clarifying questions like, "Can you help me understand what made you feel that way?" This shows a willingness to engage in meaningful dialogue rather than a desire to escalate the conflict. For example, if a friend accuses a person with ASD of being distant, the latter might say, "I'm sorry if I've seemed distant. I've been feeling overwhelmed lately and didn't mean to make you feel unimportant. Can we talk about how we can stay connected?"

This response acknowledges the other person's feelings while providing context for the behavior of the person with ASD, thus paving the way for resolution. A similar approach can be applied in professional settings when addressing workplace misunderstandings. If a colleague expresses frustration about a perceived lack of teamwork, the person with ASD might explain, "I understand that you feel I haven't been as engaged. I sometimes focus deeply on my tasks and don't always check in as much as I should. I would like to figure out how we can improve our collaboration. This can likewise help de-escalate tension and foster a discussion on solutions, not conflict. Another useful tactic is to practice structured conflict resolution strategies like taking breaks to process emotions, using written communications for clarity, or seeking mediation when necessary. With patience, self-awareness, and clear communication, conflicts can become opportunities for building stronger relationships, both personally and professionally.

Building Resilience through Communication

Improving communication skills isn't just about navigating social interactions, it's also about building resilience. Effective communication fosters more substantial relationships, reduces misunderstandings, and enhances self-confidence. For individuals with ASD, these outcomes can significantly improve their quality of life and help them engage more meaningfully in personal, professional, and social settings alike. Communication challenges may sometimes feel overwhelming, but persistence and self-awareness can transform these difficulties into opportunities for growth.

As an autistic individual strives to strengthen their communication skills, they should remember that progress takes time. They should celebrate small successes like effectively expressing a complex thought, accurately interpreting a nonverbal cue, or successfully resolving a misunderstanding. Each step forward is a testament to their resilience and determination! It's equally essential to recognize that setbacks are a natural part of the learning process. Instead of viewing them as failures, autistic adults should see them as valuable learning experiences that contribute to their growth.

Building resilience through communication involves developing the ability to adapt, stay patient, and trust in our capacity to improve. It also includes advocating for ourselves when necessary—whether that means seeking clarification in a conversation, requesting accommodations in the workplace, or expressing our needs openly in relationships. The more we practice doing these things, the more confident and empowered we'll become in our interactions with others.

Communication is a dynamic dance between words and meanings, expressions, and interpretations. By practicing active listening, clear expression, adaptability, and nonverbal cues, adults with ASD can unlock the power of effective communication. Of course, this journey requires patience, practice, and a willingness to learn from successes and setbacks—but with each conversation, each connection, and each understanding forged, the bridge of communication grows stronger. Over time, these skills will help individuals better navigate the social world and, in the process, create deeper connections and a greater sense of self-assurance. The ability to communicate effectively is a lifelong tool that, when nurtured, can open doors to new opportunities, new relationships, and personal fulfillment.

Five Key Takeaways

Explicit Teaching of Social Rules

Direct instruction on social norms—including appropriate greetings, turn-taking, and interpreting body language—can clarify expectations and reduce anxiety. Many social cues that come naturally to neurotypical individuals may not be intuitive for those with ASD, but explicit teaching can break these down into understandable steps. For example, structured lessons on how to recognize when a conversation partner is losing interest or how to interpret different tones of voice can make social interactions more predictable and manageable. Practicing these skills through guided instruction helps individuals apply them more confidently in real-life situations.

Literary and Visual Supports

Social stories, cue cards, and illustrated guides can help autistic individuals recognize and interpret social situations more effectively. These tools provide concrete, visual examples of social interactions and expectations, thereby reducing ambiguity. For instance, a visual flowchart might show how to navigate a conversation, from greeting someone to knowing when to end the discussion. Visual supports can also aid in emotional recognition by illustrating different facial expressions and their meanings. Using these resources regularly can reinforce positive communication habits and serve as references when navigating unfamiliar social settings.

Role-Playing Exercises

Practicing different conversational scenarios with a therapist, coach, or trusted friend can enhance adaptability and confidence in social interactions. Role-playing allows individuals to experience social exchanges in a controlled, low-pressure environment, making developing and refining their communications skills easier. For example, rehearsing how to introduce themselves at a networking event or practicing how to handle misunderstandings in a workplace setting can make these situations feel less intimidating when they occur in real life. Over time, exposure to different social contexts through role-playing helps individuals build flexibility and resilience in their interactions.

Encouraging Self-Advocacy

Teaching individuals with ASD to express their communication preferences can lead to more effective exchanges. Many communication breakdowns occur because others may not understand how an autistic individual best processes information. Encouraging self-advocacy—such as stating "I understand better when instructions are written down" or "I need a moment to process before responding"—empowers individuals to take control of their communication needs. When people around them understand these preferences, interactions become smoother, reducing frustration for both parties. Self-advocacy also fosters independence and confidence in both personal and professional settings.

Mindfulness and Emotional Regulation Techniques

Managing anxiety and sensory overload can enhance focus and conversational engagement. Many individuals with ASD experience heightened stress in social interactions, which can make communication more difficult. Fortunately, mindfulness techniques such as deep breathing exercises, progressive muscle relaxation, and grounding techniques can help manage anxiety and improve concentration during conversations. Emotional regulation strategies can also make interactions more manageable, like recognizing when a break is needed or using a stress reduction tool (i.e., a fidget device). By integrating mindfulness and self-regulation practices into their daily routines, autistic individuals can build greater emotional resilience and communicate more effectively in various social situations.

By applying these strategies, individuals with ASD can improve their communication skills over time, leading to more meaningful social connections, greater self-confidence, and enhanced success in both their personal and their professional lives. Communication is a skill that can be developed with patience, practice, and having the right support systems in place.

Self-Guided Activities to Improve Communication Skills

Activity 1: Practicing Active Listening

Begin by watching or listening to a conversation in a movie or TV show without pausing or rewinding it, then summarize what was said in just one or two sentences. Follow up by practicing with a friend or family member—repeat key phrases used in the conversation you heard and then ask clarifying questions about it. Reflect on your ability to maintain your attention and accurately summarize what was said, noticing how the speaker reacted when they felt heard. With regular practice, this exercise will improve your focus, memory, and responsiveness—active listening can deepen your conversations and enhance mutual understanding in social and professional interactions.

Activity 2: Nonverbal Communication Awareness

Observe a conversation in public or on TV and take notes on physical gestures and shifts in tone, then mimic these cues in front of a mirror or with a friend to increase your awareness of how others feel and look. Reflect on which cues were easiest to notice and how mirroring them influenced your level of comfort and understanding. Recognizing nonverbal signals will boost your empathy and help you respond more appropriately in social settings where spoken words don't convey the entire story. (Which is much of the time.)

Activity 3: Role-Playing Different Social Scenarios

Choose a social situation (ordering at a restaurant, attending a job interview, chatting with a friend, etc.) and role-play it with someone you trust, experimenting with different tones, speeds, and word choices to find what feels natural and effective. Afterward, reflect on which scenarios felt more or less comfortable and identify the aspects of communication (tone, gesture, vocabulary) that were hardest to navigate. Role-playing these scenarios will help you uncover your social strengths and areas for improvement, empowering you to approach real-life conversations with greater flexibility and confidence.

Activity 4: Understanding Social Cues Using Various Media Sources

Watch a TV show, movie, or sitcom known for its humor or social interactions. Whenever a sarcastic remark or joke is made, pause the scene and describe how you interpreted it, then compare your understanding with how the characters reacted to it. Think about whether the humor made sense to you immediately or not and which verbal or nonverbal cues indicated what was happening. Recognizing these subtle cues will improve your social comprehension and prepare you for real-life interactions, where humor and indirect language are frequently used to express feelings or thoughts.

Activity 5: Practicing Clear and Concise Speech

Summarize a simple daily event in one sentence, then explain a more complex idea using only three key points. You can record yourself and then listen to what you said to identify areas where you could simplify or clarify your summaries. This process will help you refine your messages and become more aware of your pacing and word choice. Reflect on whether it was challenging or not to remain concise and what strategies helped you do so. Over time, this practice will build up your confidence in your ability to communicate and will make it easier to express your thoughts in a focused and engaging way.

By engaging in these self-directed activities, you can enhance your active listening skills, better interpret social cues, and express yourself more clearly, all of which will result in you having more effective and fulfilling interactions in your daily life.

6

Self-Acceptance and Embracing Individuality

Accepting ourselves is the cornerstone of having a fulfilling and authentic life, especially for individuals with ASD. It's a process and a practice that calls for recognizing, embracing, and celebrating our unique strengths and challenges! This chapter focuses on how self-acceptance lays the groundwork for resilience and well-being, allowing autistic individuals to align their actions and decisions with personal values rather than societal expectations. By embracing their true selves, individuals can reduce the stress of conforming to external pressures and live more confidently and authentically.

Understanding Self-Acceptance

To embrace self-acceptance is to engage in a compassionate understanding of ourselves. It means recognizing both our strengths and our limitations. These limitations are not perceived as flaws to be eradicated, however, but as integral parts of our identity. This understanding becomes especially significant within the context of ASD, as societal norms often undervalue or misunderstand neurodiversity. Developing self-acceptance involves challenging these external misconceptions and fostering a positive self-image. It also requires rejecting the idea that differences are deficits and instead strives see them as variations in how individuals experience and interact with the world.

Self-acceptance does *not* mean being complacent or refusing to grow—rather, self-acceptance allows us to approach personal development from a place of self-respect. It means recognizing the difference between growth that aligns with our values and abilities and growth that stems from a desire to fit into molds created by others. True self-acceptance provides a foundation of confidence that allows us to pursue our goals without fear of being judged or feeling inadequate.

Furthermore, self-acceptance fosters resilience. When we fully accept ourselves, we become less reliant on external validation and more attuned to our own needs and aspirations. This empowers us to set boundaries, advocate for ourselves, and seek environments that nurture our strengths rather than diminish them. In a world that often demands conformity, self-acceptance is an act of quiet defiance that allows us to live authentically and unapologetically.

Challenges to Self-Acceptance and Individuality in Autistic Adults

Autistic adults frequently encounter difficulties with self-acceptance and embracing their individuality due to the cognitive challenges tied to their condition. These challenges affect their capacity to understand themselves, connect with others, and navigate a world that often emphasizes conformity to neurotypical standards. The journey toward self-acceptance is further complicated by various cognitive impairments that influence their self-perception and interactions with society.

Difficulty Understanding Mental States

One significant challenge is attributing mental states to ourselves and to others— understanding our own thoughts, emotions, and motivations is a key component of self-awareness. For individuals with ASD, though, this process can often be unclear or inconsistent. They may struggle to develop a cohesive sense of identity without fully grasping their mental states, and on top of that, difficulty interpreting the thoughts and feelings of others can lead to misunderstandings, reinforcing the belief that they're fundamentally different or incapable of forming social connections. Many autistic adults describe growing up feeling as though they were "missing a piece" that others seemed to have naturally, a perception that can lead to self-doubt and a sense of isolation. When people's social difficulties are criticized rather than being supported, they may internalize that they are somehow "wrong" instead of simply being different.

This difficulty in attributing mental states (often linked to "theory of mind" deficits) can affect various areas of life, ranging from personal relationships to professional interactions. Because they may lack the ability to intuitively understand social cues or anticipate others' reactions, individuals with ASD may face frequent social anxiety, making even everyday interactions stressful. They may also find self-reflection challenging; that in turn makes it hard to comprehend personal growth, emotions, or the effects of their actions. This difficulty can further hinder their ability to build lasting relationships, as neurotypical social expectations typically depend on implicit understanding rather than explicit communication. In many instances, autistic individuals develop compensatory strategies, such as mimicking social behaviors or relying on logical reasoning instead of instinctive empathy. As we've previously noted, while these strategies can work, they demand considerable mental energy, which can result in exhaustion and burnout.

For instance, in romantic relationships, without direct verbal communication, an autistic person may struggle to recognize when their partner is upset—if their partner relies on subtle cues like body language or tone shifts to express dissatisfaction, the person with ASD may not notice the emotional change until the issue escalates. This can lead to misunderstandings where the partner feels unheard or unappreciated while the individual with ASD feels confused and overwhelmed. However, the relationship dynamic can become more supportive and fulfilling when both partners openly communicate their expectations and needs, such as the neurotypical partner explicitly stating their feelings instead of expecting the other person to "just know." Fostering environments that encourage open communication, patience, and self-acceptance is crucial in supporting individuals with ASD as they navigate their unique social and emotional landscapes. When partners, friends, and colleagues are educated about neurodivergent communication styles, relationships can become more inclusive, strengthening mutual understanding and emotional connections.

Challenges with Emotional Expression

Struggles with emotional expression further complicate this issue. In many social situations, people expect a degree of emotional reciprocity—i.e., responding in a way that mirrors or validates the emotions of others. Individuals with ASD may not express emotions in ways that match neurotypical expectations, leading people to misunderstand their feelings and intentions. For instance, an autistic person might not outwardly show sadness when a friend shares bad news, even if they feel empathy internally. As a result, they may be seen as cold or indifferent, and over time, that perception can lead to strained relationships and social rejection. Repeated experiences like these can cause them to question their emotions and behaviors, making it increasingly difficult to embrace their natural ways of relating to others.

Additionally, the effort required to regulate and express emotions in socially acceptable ways can be overwhelming. Many individuals with ASD experience alexithymia, a condition characterized by difficulty identifying and describing one's emotions. This can make it challenging for them to communicate their feelings, even when they want to. As a result, an autistic person may suppress their emotions or rely on logical explanations rather than emotional ones, a situation that again can lead to misunderstandings in personal and professional relationships.

Sensory sensitivities can also influence emotional expression. Overstimulation from bright lights, loud noises, or crowded environments may cause distress that manifests as frustration or withdrawal that others misinterpret as a lack of concern or engagement. Without supportive people and environments that acknowledge these challenges, individuals with ASD may feel increasingly disconnected, and their feelings of isolation and self-doubt may be reinforced.

Difficulties with Planning and Organization

Challenges with executive functions—including difficulties with planning and focusing on relevant details in the environment—also shape self-perception. Many autistic adults find everyday tasks such as managing their time, organizing their responsibilities, or completing multi-step processes to be overwhelming. In workplaces or academic settings where efficiency and multitasking are often expected, these difficulties can result in feelings of inadequacy. A person who consistently forgets appointments, misplaces essential items, or struggles with shifting between tasks may feel like they are failing at life's basic expectations. This sense of failure is often reinforced by coworkers or others who become frustrated and angry with these behaviors. The resulting combination of criticism and self-criticism makes it even more difficult for an individual with ASD to learn to identify and appreciate their unique strengths—a deep focus on specialized interests, creative problem-solving abilities, etc.

Moreover, these executive function difficulties can contribute to chronic stress and anxiety. When tasks pile up and deadlines approach, autistic individuals may feel paralyzed by the pressure, unable to initiate actions even when they recognize the urgency of doing so. Procrastination, avoidance, or hyper-fixation on minor details can exacerbate these challenges and create cycles of frustration and self-doubt. In social situations, difficulty processing multiple streams of information—such as maintaining a conversation while also interpreting body language—can further increase anxiety and lead to withdrawal.

Many individuals develop coping mechanisms like rigid routines or organizational tools to address executive dysfunction. Although these strategies can be beneficial, they may also induce extra stress when they're disrupted by unexpected changes. Without adequate support and accommodations, individuals with ASD might find it challenging to build confidence in their abilities, a situation that ultimately reinforces their negative self-perceptions. Fortunately, promoting self-compassion, implementing flexible problem-solving techniques, and using structured support systems can help lessen these challenges and empower individuals to recognize their strengths instead of concentrating solely on their difficulties.

Misinterpretation of Nonverbal Cues

Another significant barrier to self-acceptance is having difficulty with interpreting nonverbal communication, particularly the communicative content of gaze. As we've already emphasized, eye contact, facial expressions, and subtle gestures carry immense weight in social interactions, yet many autistic individuals struggle to read these signals accurately. In childhood, they may have been reprimanded for not making enough eye contact or misinterpreting social cues, and that negative feedback may have made them feel socially incompetent. As adults, they might still experience anxiety in face-to-face interactions, not because they lack a desire for connection but because they feel unequipped to navigate the nuances of social engagement. These struggles can lead them to withdraw from social situations altogether, reinforcing the belief that they don't belong.

For example, an autistic adult working in a corporate environment may find it challenging to interpret a colleague's facial expressions during a meeting. If a coworker's neutral expression is mistaken for disapproval, it could trigger unnecessary anxiety and self-doubt. Similarly, they might not notice subtle signs that a conversation partner is losing interest, causing them to unintentionally dominate discussions or miss the social cues intended to wrap up a conversation. These misinterpretations can lead to workplace misunderstandings, feelings of exclusion, and self-consciousness about their social abilities. Encouraging understanding and acceptance from both individuals and those in their social environment can help alleviate the pressure to conform to neurotypical social norms and can help individuals foster more meaningful connections.

Struggles with Cooperation and Teamwork

The ability to work collaboratively with others poses another challenge. Although teamwork and collaboration are highly valued in both professional and social settings, maintaining joint attention and synchronizing behaviors with others can be difficult for someone with ASD. A person who prefers independent work or who struggles to adjust to group dynamics quickly may be seen as uncooperative or hard to work with. An employee who finds brainstorming sessions overwhelming or requires extra time to process instructions could be perceived as being disengaged rather than simply having a different cognitive approach. Over time, consistent negative feedback from coworkers or supervisors may lead autistic individuals to question their professional capabilities, making self-acceptance more challenging.

Many individuals with ASD excel when provided with clear, structured tasks involving minimal ambiguity. However, in team environments where spontaneous collaboration and quick decision-making are expected, they may struggle to keep up with changing priorities. Differences in social communication like difficulty reading between the lines or understanding implicit expectations can also lead to friction with colleagues. For example, a software developer might prefer to work independently on coding tasks but may find it challenging to participate in daily stand-up meetings or collaborative debugging sessions. If their preference for solitude is misinterpreted as a lack of interest in teamwork, they may be excluded from opportunities for career advancement.

Sensory sensitivities or anxiety related to social interactions can make group work even more exhausting. The unpredictability of teamwork, constant verbal exchanges, and unspoken social norms can create stress, causing individuals to withdraw or mask their challenges. When workplace environments fail to recognize these difficulties and provide accommodations (e.g., offering written instructions, alternative communication methods, or structured roles within teams), it can reinforce feelings of alienation. Promoting neurodiversity-friendly workplace practices—including flexibility with collaboration styles—can help autistic individuals feel valued and confident in their professional abilities.

Difficulty Navigating Social Norms

Understanding and interpreting complex social information is another factor that contributes to difficulties with self-acceptance. Society is full of unspoken rules, expectations, and shifting social dynamics that individuals with ASD may find hard to decode—something as seemingly simple as knowing when to end a conversation, recognizing sarcasm, or grasping hierarchical workplace relationships can be confusing and exhausting. These challenges often lead to missteps or awkward interactions, resulting in self-consciousness and anxiety. A person who has repeatedly faced social rejection or embarrassment may develop a profound sense of insecurity; they may question their worth and struggle to feel comfortable in their own skin.

For instance, an autistic adult working in an office environment might struggle to navigate informal workplace interactions. They may not recognize when small talk is expected and may instead jump into a task-related discussion, causing coworkers to perceive them as abrupt or unfriendly. Similarly, they may misinterpret a sarcastic remark as a serious statement and then respond in a way that seems out of place. These miscommunications can create tension and frustration, making social interactions feel like a minefield of potential mistakes.

The fluid nature of social hierarchies and relationships introduces another layer of complexity. While neurotypical individuals often rely on intuition to modify their behavior based on context, a person with ASD may need clear guidance to understand when casual interactions are suitable and when more formal conduct is necessary. This uncertainty can lead them to avoid social situations altogether, which reinforces feelings of isolation.

Ongoing social challenges can prompt autistic individuals to question their ability to form meaningful relationships, another factor that leads to a negative self-image. To tackle this issue, structured social skills training, clear communication from peers, and increased societal awareness of neurodiverse perspectives can create a more inclusive environment, allowing individuals with ASD to feel accepted and appreciated for their unique ways of thinking and interacting.

Inconsistencies in Social Learning

Compounding all these challenges is the difficulty of generalizing and applying social learning across various contexts. Many autistic individuals work diligently to learn social skills in one setting, only to discover that those same strategies don't transfer well to another—a person who masters polite conversation at work may find it challenging to apply the same skills in casual social situations, for example. The unpredictability of social interactions can make them feel as if they're continually performing rather than engaging naturally, leading them to feel exhausted and disconnected from their true selves. This struggle to maintain consistency in social behavior can make it difficult for them to develop confidence in their interactions and can reinforce a cycle of self-doubt.

An instance of this is a person with ASD who practices formal workplace etiquette like maintaining professional eye contact, using structured speech patterns, and sticking to predefined conversation topics. They may not realize that in a relaxed social setting, others will perceive these behaviors as being rigid or overly formal. When trying to engage in informal

conversations, they may struggle with the fluidity of small talk and end up feeling awkward or excluded. If they attempt to compensate by mimicking the conversational styles of others, they may feel forced or unnatural, a mindset that will contribute to their social fatigue.

Additionally, different social environments often come with unspoken expectations that shift rapidly. People who learn that it's polite to wait for their turn to speak in professional meetings may struggle in fast-paced social settings where interruptions and quick exchanges are common. That may make their social engagements feel more like intellectual exercises than natural, enjoyable experiences.

This inconsistency can erode self-confidence, as autistic individuals may believe they can never fully master social interactions. Rather than feeling empowered by the skills they *have* learned, they may grow increasingly anxious, fearing that every new social setting presents a fresh set of unspoken rules to decipher. Encouraging self-acceptance and helping individuals recognize that social learning is a lifelong process—*not* an expectation of perfection!—can reduce this pressure and promote greater self-confidence.

Impact on Self-Acceptance and Individuality

All of these cognitive challenges make it difficult to embrace individuality. When society consistently questions or misunderstands a person's natural way of thinking, feeling, and interacting, they may feel pressured to conform or conceal their true selves. As we noted earlier, many autistic adults develop masking behaviors—consciously imitating neurotypical social norms to fit in—at a significant emotional and mental cost. The effort required to maintain this façade can be exhausting and can lead to burnout, anxiety, and a sense of living inauthentically. When individuals with ASD feel they must suppress their true selves to gain acceptance, it becomes nearly impossible for them to fully embrace and celebrate their individuality.

For instance, a young professional with ASD may spend their entire workday masking: modulating their tone, suppressing stimming behaviors, and forcing themselves to engage in small talk. By the end of the day, they may feel drained and overstimulated and may struggle to find the energy to participate in activities they actually enjoy. Over time, this constant effort can lead to burnout, a state of extreme exhaustion that affects a person's mental, emotional, and even physical health.

Encouraging self-acceptance requires creating environments where individuals with autism feel safe being themselves without fear of judgment. When society values neurodiversity and recognizes that different ways of thinking and interacting are valid, autistic people can begin to unmask, embrace their true selves, and thrive authentically.

The Journey toward Self-Acceptance

The journey toward self-acceptance is not linear. It requires patience, self-compassion, and often the unlearning of negative self-perceptions that have developed as a result of years of external criticism. Many individuals with ASD grow up feeling different without understanding why. This lack of understanding can lead to feelings of isolation, self-doubt, and even self-rejection. However, with the proper support and resources, individuals can learn to see their differences as strengths rather than weaknesses.

One effective way to promote self-acceptance is through education and awareness. Learning about ASD, its characteristics, and the experiences of others on the spectrum can offer clarity and validation. When individuals realize that others share similar experiences, they feel less alone.

Another critical aspect of this journey is self-compassion. Instead of viewing social challenges, sensory sensitivities, or differences in communication as deficits, individuals can reframe these traits as natural variations of the human experience. Accepting that struggles are part of life—just as they are for neurotypical individuals—helps cultivate a more forgiving and affirming self-perspective.

The Role of Personal Narratives

Each person's narrative (i.e., their story about who they are) shapes their self-perception. For autistic adults, these narratives may have been influenced by years of external misunderstandings, exclusion, or unrealistic expectations. Rewriting these narratives to reflect a positive and authentic self-image can be transformative.

Reframing One's Narrative

Consider Cyndi, a 34-year-old woman diagnosed with ASD in adulthood. For years, she believed that her struggles in social situations and her need for structured routines made her "less-than." She often felt out of place at work, where small talk and networking seemed effortless for others but exhausting for her. It wasn't until she began to learn about neurodiversity that she realized these traits were not flaws but differences. She started to see her ability to focus intensely on her interests and her meticulous approach to problem-solving as assets. She also recognized that her preference for direct communication and honesty—which had once led to misunderstandings—could be strengths in the right environment.

Reframing her narrative allowed Cyndi to embrace her individuality and recognize her worth. She sought out supportive communities where she could connect with others who shared similar experiences, finding that those connections helped make her feel less alone. She also began advocating for accommodations at work (i.e., written instructions instead of verbal ones) that helped her thrive. Over time, Cyndi stopped seeing herself as someone who needed to "fix" herself and instead as someone who had valuable contributions to offer. This shift in perspective was transformative! It allowed her to approach life with more confidence and self-acceptance.

Aligning Actions with Values

One of the most empowering aspects of self-acceptance is aligning actions with personal values. For individuals with ASD, this may involve prioritizing sensory-friendly environments, pursuing passions that bring joy, or establishing boundaries that safeguard their well-being. It also includes recognizing that their needs are not inconveniences but vital elements of their identity that warrant respect and accommodation.

By embracing their values, individuals can gain the confidence to reject expectations or roles that don't align with their authentic selves. Todd, a 40-year-old with ASD who worked in IT, realized that his job required frequent participation in large meetings. Unfortunately, the constant pressure to contribute verbally in high-energy discussions left him mentally exhausted, drained his energy, and led to burnout. It became difficult to focus on the work he enjoyed. After considerable reflection and the assistance of a coach, he advocated for a role focused more on independent work. This allowed him to thrive without compromising his mental health. With accommodations such as asynchronous communications and written task assignments, he became more productive and felt more in control of his professional life.

Aligning actions with values is an act of self-respect. It conveys to ourselves and others that our needs, interests, and boundaries are valid. For autistic individuals, this alignment often means rejecting societal standards that undervalue neurodivergent traits and choosing to create lives that reflect their unique priorities. This shift enables them to transition from merely surviving in a world designed for neurotypical individuals to truly thriving in environments that support their strengths and well-being.

The Role of Self-Advocacy

Self-advocacy is an essential skill for aligning actions with internal values—effectively communicating needs and boundaries enables individuals to create environments where they can succeed. This may involve requesting workplace accommodations, establishing social boundaries, or pursuing educational opportunities that match their learning style. Self-advocacy is a practice that improves with time and experience, and it too is a vital tool for achieving self-acceptance.

For many individuals with ASD, self-advocacy begins with self-awareness: understanding their sensory preferences, communication style, and cognitive needs. This awareness empowers them to express their needs for comfort and success. In a workplace setting, for instance, this might involve requesting noise-canceling headphones to reduce auditory distractions or asking for written instructions rather than verbal ones for greater clarity. In social settings, self-advocacy may include explaining the need for direct communications or preferring one-on-one conversations over large group gatherings.

However, self-advocacy can be challenging. It requires confidence, persistence, and navigating conversations where others may not immediately understand or accommodate requests. Many autistic individuals fear being perceived as demanding, a concern that leads them to suppress their needs instead of voicing them. Overcoming this fear often involves building supportive networks, practicing clear communication, and reframing self-advocacy as an act of self-respect rather than an inconvenience. As people with ASD gain experience in advocating for themselves, they become more empowered to shape their lives in ways that align with their values, fostering greater confidence, independence, and self-acceptance.

The Connection Between Self-Acceptance and Resilience

Resilience is the ability to adapt and thrive in the face of challenges, and it's deeply rooted in self-acceptance. When individuals accept themselves, they're better equipped to cope with adversity without internalizing self-doubt or shame. Rather than viewing difficulties as personal failures, they can see them as part of the learning process, allowing for growth and self-improvement.

Lisa, a young adult with ASD, struggled with feelings of inadequacy due to her challenges with maintaining eye contact and understanding social cues. She often felt isolated, convinced she was failing at basic social interactions that seemed effortless to others. But after working with a counselor who specialized in neurodiversity, she began to recognize that her social style was simply different, not defective. By embracing this perspective, Lisa discovered new ways to connect with others, such as through written communications and shared interests. She realized that deep, meaningful friendships could thrive within online communities, structured group activities, and one-on-one conversations where she felt more comfortable. Her self-acceptance became the foundation of her resilience and empowered her to confidently approach social situations.

Resilience also means recognizing that setbacks and mistakes are part of personal growth. Self-acceptance offers the grace to learn from these experiences without self-judgment, and it enables individuals to view challenges as opportunities for growth rather than as proof of inadequacy. For example, now if Lisa misinterprets a social cue or struggles in a group conversation, she no longer sees it as a failure. Instead, she reflects on the experience, learns from it, and adjusts her approach without engaging in harsh self-criticism.

Resilience is also strengthened when autistic adults surround themselves with supportive individuals who appreciate neurodivergent strengths rather than focus on perceived deficits. When individuals cultivate relationships where they feel accepted and valued, they build strong emotional foundations that reinforce their ability to face obstacles with confidence. Ultimately, resilience is not about avoiding challenges but about developing the inner strength to persevere. Self-acceptance provides the foundation for this process—it allows individuals with ASD to manage life's complexities with a sense of self-worth, adaptability, and optimism for the future.

One of the most liberating aspects of self-acceptance is the realization that societal expectations don't define our worth. For autistic adults, societal pressures to conform can be overwhelming and can often lead to feelings of alienation. Embracing individuality means rejecting these pressures in favor of authenticity.

Cultivating authenticity may involve small but significant acts, such as wearing noise canceling headphones in public without fear of judgment or openly discussing sensory needs with friends and colleagues. It might mean redefining success on an individual's own terms, such as them pursuing a career that aligns with their true passions as opposed to what society deems to be prestigious.

Rejecting societal expectations requires courage. It involves standing firm in the face of external criticism or misunderstanding. Yet as many individuals with ASD (and without) have discovered, living life in concert with their values is far more rewarding than conforming to norms that diminish their sense of self.

The Power of Community and Support

A key point we've expressed throughout this chapter is that self-acceptance is not a solitary journey. Community and support play crucial roles in fostering a positive self-image. Connecting with others who share similar experiences can offer validation and encouragement, helping autistic individuals recognize that they are not alone in their struggles and triumphs. Understanding that others have faced and overcome similar challenges can be profoundly reassuring and empowering!

Support groups, online forums, and advocacy organizations offer spaces where individuals with ASD can share their stories and learn from one another. These communities reinforce the value of neurodivergent traits and the belief that everyone deserves respect and understanding. Engaging with others in these spaces can also provide practical strategies for self-advocacy, coping mechanisms, and insights into personal growth.

Family, friends, and professionals play a crucial role in this process. Supportive relationships offer the safety and encouragement that individuals need to explore their identity and embrace their individuality. When others accept and celebrate a person's uniqueness, it becomes easier for that individual to do the same. A workplace that accommodates various communication styles, a teacher who recognizes and nurtures a student's strengths, or a friend who respects sensory needs all contribute to an environment where self-acceptance flourishes, enabling individuals to display authenticity and confidence.

Embracing Authenticity: The Power of Self-Acceptance

Self-acceptance is both a journey and a destination. For autistic adults, it's key to embracing their individuality, building resilience, and living authentically. This process involves rewriting personal narratives, aligning actions with values, rejecting societal pressures, and seeking community support. Above all, the process of individuals accepting themselves emphasizes that *everyone*—regardless of their differences—deserves respect, dignity, and self-love.

In contrast, self-acceptance is *not* about changing ourselves to fit into societal norms. It's about recognizing our strengths and embracing a unique way of thinking! Self-acceptance allows all of us to move beyond self-doubt and take pride in our identity. By fostering environments that support neurodiversity, society can empower autistic individuals to thrive rather than struggle to conform.

In the words of Dr. Temple Grandin, a prominent advocate for autism awareness: "The world needs all kinds of minds." Embracing this diversity begins with self-acceptance—an act that transforms individuals and enriches the world around them. When individuals with ASD are encouraged to be their authentic selves, their creativity, intelligence, and passion shine through, benefiting both their well-being and the broader community.

Five Key Takeaways

Self-Awareness Builds Self-Acceptance

Understanding our strengths, challenges, and unique traits is essential for self-acceptance. Instead of viewing differences as deficits, autistic individuals can learn to appreciate their distinctive abilities—deep focus, creativity, honesty—as being valuable assets. Self-awareness builds confidence and encourages individuals to embrace their authentic self.

Reframing Negative Narratives

Many adults with ASD have internalized negative messages from their past experiences. Transforming these narratives from "I am not good enough" to "I am different, and that's okay" can be life-changing. Engaging in positive self-talk, learning about neurodiversity, and seeking supportive communities can help individuals reframe these narratives into self-affirming perspectives.

Setting boundaries Supports Authenticity

Self-acceptance entails establishing boundaries that respect individual needs. Whether it involves avoiding overwhelming social situations, selecting sensory-friendly environments, or declining tasks that may lead to burnout, honoring personal limits strengthens self-worth. When individuals embrace their individuality, that means they prioritize their well-being over societal expectations, an approach that results in a more fulfilling and balanced life.

Authenticity over Masking

Many autistic individuals engage in masking—that is, they adapt their behaviors to align with social norms. While this may promote acceptance in certain situations, excessive masking can be exhausting and undermine their self-identity. In contrast, embracing their natural methods of communication and interaction fosters genuine connections and a stronger sense of self-worth.

Supportive Communities Foster Growth

Finding communities that value neurodiversity provides validation and encouragement. Connecting with like-minded individuals through support groups, online forums, and friendships reduces isolation. Self-acceptance becomes less formidable when individuals are surrounded by people who appreciate different perspectives and celebrate strengths rather than focusing on perceived weaknesses.

Self-Guided Activities to Improve Self-Acceptance

Activity 1: Practicing Daily Self-Affirmations

Each morning, write down or say aloud three positive affirmations about yourself. Focus on strengths such as creativity, honesty, or perseverance. Repeating these affirmations will help you challenge your negative self-perceptions and will reinforce your self-worth over time. Keep a journal of your affirmations to track your progress and notice patterns in your self-perception.

Activity 2: Reframing Negative Thoughts

When you think negatively about a challenge or trait, pause and reframe it. For example, if you think, "I'm bad at socializing," reframe it to "I socialize differently, and that's okay." Practice writing down these shifts in a journal to develop a habit of self-compassion.

Activity 3: Creating a Personal Values List

Write down ten things that matter most to you (i.e., honesty, independence, creativity, kindness), then reflect on how your daily actions align with these values. This exercise will help you clarify what matters most to you, allowing you to focus on living authentically rather than conforming to external expectations.

Activity 4: Practicing Saying no

Identify situations where you feel pressured to conform or overextend yourself. Practice setting boundaries by politely declining requests that don't align with your needs or well-being. Roleplay these scenarios or write scripts for saying no in a comfortable way. This will reinforce your right to prioritize yourself.

Activity 5: Finding and Celebrating your Strengths

Make a list of things you enjoy and excel at, no matter how small. These strengths could include problem-solving, memorizing facts, expressing yourself artistically, or analytical thinking. Find ways to incorporate these strengths into your daily life! Recognizing and celebrating your personal achievements will make you more confident and will reinforce a positive self-identity.

By consistently practicing these exercises, you can cultivate self-acceptance, embrace your uniqueness, and lead a more authentic and fulfilling life.

7

Building Meaningful Connections: The Power of Relationships

Relationships form the foundation of human resilience—they provide emotional grounding, support, and a sense of belonging For autistic adults, building and maintaining these connections can be both rewarding and challenging. But while social norms and expectations may sometimes feel overwhelming or unclear, meaningful relationships offer essential emotional and psychological benefits. These relationships—whether with family, friends, colleagues, or community—are transformative in fostering personal growth and wellbeing. A strong support system not only helps individuals navigate daily challenges, it also promotes a sense of security and allows them to express themselves authentically without fear of judgment.

For individuals with ASD, the journey to forming meaningful connections often begins with understanding their unique social needs and boundaries. Each person experiences relationships differently—there is no single way to connect with others. Some may find comfort in structured group settings based on shared interests, while others may prefer one-on-one interactions or online communities where communication is more predictable. These connections don't always have to follow traditional pathways—they can be shaped to celebrate individuality, respect comfort levels, and embrace neurodivergent perspectives. Environments that support different communication styles and interaction preferences allow autistic individuals to cultivate fulfilling and lasting relationships.

The Foundations of Relationships

At its core, every relationship thrives on mutual understanding, trust, and compassion. However, as we've repeatedly emphasized, for individuals with ASD, the complexities of communication and social interactions can feel overwhelming. Social cues, unspoken expectations, and emotional reciprocity can seem difficult to decipher. But authentic relationships are not about perfection, they're about connection.

Lucy is a 28-year-old professional with ASD who has often found traditional social situations be to overwhelming. However, she discovered deep connections in smaller, structured group settings centered around shared interests, like a book club focused on science-fiction novels. By engaging in conversations about topics she loves, Lucy has built meaningful and manageable relationships.

The key to successful relationships lies in aligning how interactions occur with respect to individual strengths and preferences. For autistic adults, this might mean favoring smaller gatherings over large parties, written communications over phone calls, or shared activities over casual conversations. Authenticity is key—relationships flourish when they align with an individual's true self.

Understanding Mental States: A Core Challenge

As we've mentioned, one of the most fundamental challenges individuals with ASD face is the ability to attribute mental states (beliefs, intentions, desires, emotions) to themselves and others. This skill as noted earlier, is often referred to as "theory of mind." It allows people to interpret social situations and predict the behavior of others. However, for many autistic individuals, understanding that others may have different thoughts, motivations, or perspectives can be difficult.

We previously met Dan, a 27-year-old man with ASD, who works as a software engineer. During a meeting, his manager became frustrated with a delayed project, started to sigh heavily and rub his temples. Dan was unaware that these nonverbal cues were indicating stress, so he continued discussing technical details, not realizing that his manager was unresponsive. His difficulty in recognizing the mental states of others has led to misunderstandings in the workplace and has made it challenging for him to navigate social expectations.

Like Dan, because they may not recognize that others may think differently, individuals with ASD might struggle with social reciprocity, which can lead to unintentional conflicts. Learning to employ strategies like asking clarifying questions ("Are you feeling frustrated?") or relying on structured social scripts can help individuals bridge this gap. Additionally, neurotypical peers can promote inclusivity by being more explicit about their emotions and expectations.

Emotional Reciprocity: Navigating Joint Attention of Emotion

Emotional reciprocity, or the ability to show an emotional response that aligns with another person's mental state, is another area where autistic individuals may have difficulties. Emotional reciprocity enables individuals to forge meaningful social connections by responding to the feelings of others in a manner that promotes bonding, and when this is a challenge for someone, that can result in them being perceived as insensitive even when they have no intent to be unkind.

We met Jill in Chapter Three. She valued her friendships, especially with her roommate Lisa. However, she found it challenging to reflect the emotions of those around her. When Lisa shared exciting news about a promotion, Jill responded with a simple "That's good"; she lacked the enthusiasm typically expected in such moments. Because she had been anticipating a more animated reaction, Lisa felt as though she was being dismissed even though Jill deeply cared about her and their relationship. Still, because Jill didn't instinctively align her emotional expressions with those of other people, she sometimes appeared to be uninterested in their lives. Upon recognizing this challenge, Jill explained to her friends that her reactions might not always align with her internal feelings, and in return, her friends learned to appreciate her supportive nature beyond conventional expressions of enthusiasm. Understanding these differences helped foster acceptance and smoother interactions between everyone involved. As Jill learned that support from understanding colleagues, friends, and loved ones could help alleviate some of the burdens she had been experiencing in her relationships, making emotional connections became more accessible and less overwhelming.

Attention to Detail and Environmental Awareness

Another challenge for individuals with ASD is balancing their attention to detail with an awareness of their broader social environment—many individuals excel at focusing on specific elements but struggle to integrate multiple sources of information simultaneously. This can lead to difficulties in social and professional settings, where the ability to shift attention between details and overall context is essential.

Justin, an engineer with ASD we discussed in Chapter Four, meticulously reviewed every specification in a design document but struggled to prioritize which aspects required his immediate attention. During meetings, he would become hyper-focused on minor technicalities and miss the larger discussion about project deadlines. His colleagues grew frustrated because they felt he wasn't considering their broader concerns.

Through coaching and self-awareness exercises, Justin learned how to prioritize important information by using visual reminders and structured outlines. He also expressed his need for clear instructions, enabling his colleagues to assist him in balancing details with the overall context. This adaptation allowed him to work more effectively in a team and also reduced misunderstandings and enhanced collaborative efforts.

Understanding the Communicative Content of Gaze

Eye contact and gaze direction are critical in nonverbal communication—both convey emotions, attention, and intentions. However, many autistic individuals find it difficult to interpret these cues, leading to challenges in social interactions. Misreading or avoiding eye contact can cause others to misinterpret their intentions and can create barriers to connection. Gina, a college student with ASD, avoided eye contact because it felt overwhelming— during conversations, she would look at the floor or to the side, which made her classmates assume she was disinterested or not paying attention. As a result, they engaged with her less, limiting her opportunities for social interaction.

With the assistance of a therapist, Gina learned that while maintaining eye contact could be uncomfortable, occasionally glancing at a speaker's nose or forehead could help her seem engaged without causing her distress. Gina also informed her peers about her communication preferences, which promoted better mutual understanding for everyone. By finding alternative ways to show her engagement—such as nodding or summarizing what was said—Gina strengthened her connections with classmates while staying true to her comfort levels.

Cooperative Behavior and Joint Attention

Working cooperatively with others involves sharing attention, taking turns, and collaborating effectively. This can often be a challenge for individuals with ASD, who may find it difficult to navigate the fluid give-and-take of group dynamics. Their natural inclination toward independent thinking is sometimes misunderstood as a lack of willingness to cooperate.

James, a graphic designer with ASD introduced in Chapter One, enjoyed problem-solving but preferred working alone. He fully focused on his assigned tasks during team projects, often without checking in with his colleagues. When his team discussed ideas, he contributed only when directly asked, a tendency that led others to assume he was disengaged. However, James valued teamwork—he was simply struggling with the unstructured nature of group discussions. Through practice and mentorship, James adopted small strategies like setting reminders to ask teammates about their progress and participating in brief check-ins. His team also adapted by providing clearer roles and more structured collaborative methods. By finding ways to engage that aligned with his strengths, James enhanced his teamwork skills while maintaining his preferred working style.

Processing and Analyzing Social Information

Interpreting and analyzing social information necessitates grasping context, tone, and implicit meanings. Many autistic individuals find these nuances challenging, which can result in difficulties during conversations, jokes, and indirect communications.

Ruth, a marketing professional with ASD, took language literally and struggled with sarcasm and implied meaning. When a colleague jokingly said, "Great, another meeting! Just what we needed!", Ruth assumed her colleague was being genuinely enthusiastic. Later, she realized she had misinterpreted the statement. That led to her being confused about her colleagues' attitudes.

To navigate these situations, Ruth began using context clues and asked clarifying questions when she was unsure of what was actually being expressed.

She also informed close colleagues about her preference for direct communication. Over time, her team members became more mindful of their phrasing, while Ruth became more comfortable with recognizing tone and intent. This mutual adaptation allowed for more precise, more effective communication in the workplace.

Individuals with ASD encounter unique challenges in forming and sustaining relationships due to differences in mental state attribution, emotional reciprocity, environmental focus, gaze communication, cooperation, and social information processing. By recognizing these challenges and utilizing strategies to address them, both autistic individuals and their neurotypical peers can foster more inclusive and meaningful connections. With patience, education, and mutual respect, the path toward deeper relationships becomes more attainable and rewarding for everyone involved.

Navigating Challenges

Relationships come with challenges, especially when individuals must navigate misunderstandings or differing expectations. But even though autistic adults might encounter difficulties in interpreting others' emotions or articulating their own needs, these challenges aren't insurmountable. Practicing clear and direct communication can alleviate misunderstandings, for one thing—explaining preferences, setting boundaries, or expressing appreciation in explicit terms can go a long way toward building trust. Neurotypical friends or partners should also educate themselves about ASD so that they too can foster mutual understanding and empathy.

Gary, an engineer with ASD, found workplace interactions challenging. By proactively sharing his need for clear instructions and occasional quiet time, he helped his colleagues understand his working style. In response, they adjusted their approach, creating a more inclusive and supportive environment. This underscores how transparency can transform potential friction into a foundation for stronger relationships. In a similar way, Gloria, a woman with ASD, learned to express her preference for visual schedules to her boyfriend in order to minimize misunderstandings at home. By explaining that written plans made her feel secure and organized, she reduced her anxiety and encouraged a deeper sense of teamwork in their relationship.

The Role of Social Skills

While relationships thrive on authenticity, specific social skills can make interactions smoother and more fulfilling. For autistic adults, learning these skills is often a matter of practice and exposure rather than innate intuition. Fortunately, social interactions, though complex, can become more manageable with effort and guidance.

Strategies such as active listening, maintaining comfortable eye contact, and asking open-ended questions can enhance interactions. Role-playing scenarios with trusted friends, family members, or therapists can also build confidence in navigating social situations. For example, rehearsing how to approach a new acquaintance at a hobby group can transform initial anxiety into a positive experience.

It's important to emphasize that developing social skills does *not* mean suppressing individuality. Instead, it's about finding a balance that allows for self-expression while engaging meaningfully with others. The goal is not conformity but communication—in other words, individuals are enabled to connect in authentic and empowering ways.

Nurturing Compassion in Relationships

Compassion is the glue that holds relationships together. For adults with ASD, compassion can take many forms, from showing patience in understanding a friend's perspective to offering support during difficult times. Compassion is not a one-way street; it's equally vital to receive compassion from others.

Meg, a 42-year-old teacher diagnosed with ASD, exemplifies the power of compassion in relationships. She once shared how her close friends helped her navigate a challenging time. "Their kindness wasn't about fixing my struggles," she explained. "It was about being there and reminding me I wasn't alone." Her experience highlights that compassion, no matter how small, fosters deep bonds and mutual respect.

Compassion also includes having compassion for ourselves. Autistic adults frequently face internalized stigma and self-doubt, making self-compassion an essential tool for fostering resilience. Learning how to forgive themselves for mistakes, creating room for growth, and recognizing their strengths can establish a foundation for healthier and more fulfilling relationships.

The Importance of Social Support Networks

Social support networks are vital for resilience and personal growth. These networks include family members, friends, mentors, colleagues, and online communities. For individuals with ASD, finding or creating supportive spaces where they can be themselves is particularly important.

Online communities, for example, have become a haven for many adults with ASD.

These platforms allow individuals to connect based on shared interests or experiences, bypassing the pressures of face-to-face interactions. Within such communities, individuals can form meaningful and empowering relationships through forums, social media groups, and virtual meetups.

In-person support networks also play a crucial role. Whether it's a close-knit group of friends, supportive colleagues, or an autism advocacy group, these networks provide validation, understanding, and encouragement. Online and offline support groups can also help individuals share strategies and experiences, which fosters a sense of solidarity.

Relationships as a Buffer against Stress

The power of relationships lies in their ability to buffer against stress. For autistic individuals, the world can sometimes feel overwhelming—sensory sensitivities, social pressures, and everyday challenges can add up. Relationships offer a safe space to decompress, share experiences, and recharge.

Stan, a man with ASD, found his workplace demanding. His small circle of friends outside of work became his sanctuary—they provided him with a space to vent, laugh, and feel understood. This support helped him maintain his mental health and resilience, even during tough times.

Research supports the idea that strong social connections can reduce the impact of stress on mental and physical health. The mere presence of a trusted friend or supportive partner can lower cortisol levels, decrease anxiety, and promote a sense of stability. For adults with ASD, this buffering effect can be a powerful tool in managing daily challenges.

Building a Sense of Belonging

Belonging is a universal human need, and for autistic adults, fostering this sense of belonging requires both self-acceptance and external validation. By seeking out relationships that celebrate differences and promote inclusivity, individuals can create spaces where they feel valued.

This might involve joining advocacy groups, participating in hobbies with like-minded individuals, or simply spending time with people who embrace neurodiversity. The goal is not to fit into preconceived molds but to form connections that reflect and honor our true selves. For instance, Cal, a middle-aged music enthusiast with ASD, joined a local choir where his love for singing found resonance. He built friendships and discovered a sense of purpose and pride in contributing to group performances. This sense of belonging became a source of joy and self-expression.

Romantic Relationships and ASD

Romantic relationships can be both fulfilling and complex for individuals with ASD. While the desire for love, connection, and companionship is universal, those on the spectrum may encounter unique challenges in navigating the emotional, social, and communicative facets of intimate relationships. The key to success lies in understanding these challenges, fostering clear communication, and embracing relationships that align with their needs and strengths. One of the most significant challenges in romantic relationships for autistic individuals is understanding and attributing mental states to their partners. This difficulty in perceiving a partner's thoughts, emotions, or unspoken expectations can lead to misunderstandings. For example, John whom we described in Chapter One, struggled to recognize when his wife Sarah was upset unless she stated so clearly. While Sarah expected him to pick up on subtle cues like body language or changes in tone, John required more direct communication. Over time, they worked together to bridge this gap by openly discussing feelings and setting clear expectations for emotional support.

Another challenge involves displaying emotional reactions appropriate to a partner's state. (This is often called joint attention of emotion.) Expressing emotions in a way that aligns with a partner's expectations can be difficult. Some autistic individuals may express love and affection unconventionally, such as through shared interests or practical actions rather than verbal affirmations or physical affection. Nicholas, a man with ASD, showed his love by fixing small things around the house and organizing his partner's schedule rather than offering frequent compliments. While his intentions were caring, his partner initially misunderstood his actions as him being emotionally distant. Through open discussions, however, they found a balance where he could communicate love in his way while also trying to meet his partner's needs. Planning and attending to relevant details in the environment is another area that can impact romantic relationships. Social and emotional nuances that seem automatic to neurotypical partners may require deliberate effort for autistic individuals. This includes planning dates, remembering anniversaries, or recognizing the importance of spontaneous gestures. For example, Gloria, mentioned earlier in this chapter, preferred visual schedules., loved structured routines and became anxious when her boyfriend made last-minute changes to their plans. Understanding this, her boyfriend learned to communicate changes in advance, while she worked on embracing some flexibility. This joint effort of mutually respecting each other's needs allowed their relationship to thrive.

Understanding the communicative content of gaze can also play a role in romantic relationships. Eye contact is often viewed as a sign of emotional connection, but for many individuals with ASD, maintaining eye contact can be uncomfortable or distracting. Partners can misinterpret this difference as a lack of interest or engagement. Instead of forcing eye contact, couples can find alternative ways to express connection, such as using verbal affirmations, touch, or shared activities. One married couple developed a system where the husband would squeeze his wife's hand when he wanted to express that he was actively listening, making their communication more comfortable and reassuring.

Working cooperatively in a relationship (known as joint attention of behavior) requires navigating shared responsibilities, decision-making, and compromise. This can be challenging for autistic individuals who prefer structured routines or have specific ways of doing things. Conflict may arise if a partner expects a more spontaneous or adaptable approach. Understanding and discussing these differences early on helps couples create systems that work for both individuals. For instance, one couple developed a structured approach to household tasks using a shared digital calendar. This tactic ensured clarity and reduced stress.

Comprehending and analyzing social information in the relationship environment is also crucial. Many unwritten rules exist in romantic relationships, from knowing when to initiate conversations to recognizing when a partner needs emotional support. Because individuals with ASD often process social situations differently, having direct and open discussions about expectations can prevent miscommunications. For instance, Gloria sometimes struggled with knowing when her boyfriend needed emotional reassurance versus space. Establishing a system where her partner could express his needs explicitly made their relationship more substantial and more supportive.

Romantic relationships require effort, compromise, and understanding from both partners. For autistic individuals, these relationships can flourish when both partners are willing to communicate openly, respect neurodivergent differences, and find meaningful ways to express love and connection.

Parent-Child Relationships and ASD

Parent-child relationships can be challenging for individuals with ASD. Whether an individual with ASD is the parent or the child, the unique ways in which they process emotions, social cues, and communication can shape the dynamics of their bond. Understanding these challenges and working to build trust, clear communication, and mutual understanding can help strengthen these relationships in a way that honors neurodivergence.

One of the key challenges in parent-child relationships for autistic individuals is attributing mental states to others. A parent with ASD might struggle to recognize their child's emotional needs, especially when those needs are expressed subtly and nonverbally. For instance, Jonah, a father with ASD, found it challenging to interpret his son's need for comfort after a tough day. Instead of seeing his son's withdrawn behavior as a sign of sadness, Jonah assumed he wanted to be left alone. Through discussions with his wife, however, Jonah learned to ask direct questions about his son's feelings instead of relying solely on nonverbal cues. This approach helped him provide the support his child needed.

Another challenge is displaying emotional reactions that are appropriate to another person's state (also known as joint attention of emotion). Some autistic individuals may struggle to mirror their child's excitement, frustration, or sadness in the expected way. For instance, a mother with ASD might not instinctively match her child's enthusiasm about a new toy or an achievement at school. That doesn't mean she isn't proud or happy, it simply means her emotional expression may appear different. One way to bridge this gap is for individuals to explicitly use words to express emotions, such as saying "I'm happy for you" even if they aren't outwardly displaying their excitement in the way the child might be anticipating.

Planning and paying attention to relevant details is another important aspect of parenting. Many parents with ASD thrive on routine and structure, but children—especially young ones!— can be unpredictable. Adjusting to schedule changes, responding to a child's shifting needs, and handling daily parenting tasks can be overwhelming. Sonya, a mother with ASD, found mornings to be extremely stressful as her daughter would often change her mind about what she wanted to wear or eat. To alleviate her anxiety, this mother created a structured morning checklist with limited choices for her daughter. That helped both mother and daughter feel more in control.

Understanding the communicative aspects of gaze can also be important in parent-child relationships. Some autistic parents may not use much eye contact, which children might misinterpret as disengagement. However, parents can demonstrate attention in other ways: through verbal affirmations, physical closeness, or participating in shared activities. A father who struggles with eye contact might express his love by reading to his child every night or listening attentively when they speak, for example.

Parenting involves collaborating in a relationship and paying joint attention of behavior. It involves a partnership between the caregiver and the child, and it requires cooperation, patience, and compromise. Some parents with ASD may find it difficult to manage moments when children seek immediate attention, interrupt conversations, or exhibit unpredictable behaviors. Developing strategies like creating visual schedules or establishing clear rules for transitions can help maintain structure while ensuring that children feel heard and valued. Ultimately, understanding and analyzing social information within the parenting context can influence how an autistic parent engages with their child. The social expectations of parenting—such as knowing when to be firm, when to show nurturing, or how to partake in imaginative play—may not always come instinctively. For one thing, parents on the spectrum often find comfort in structured activities instead of unstructured play. Jonah enjoyed assembling LEGO sets with his son because doing so gave them a clear, goal-oriented activity they could share. Discovering ways to bond through shared interests can strengthen parent-child relationships, honoring both the child's needs and the parent's needs.

Parenting is a journey of continuous learning and adaptation. For individuals with ASD, embracing strategies that align with their strengths *and* foster open communication with their children can help build strong, loving relationships. The goal is not to conform to neurotypical parenting expectations but to cultivate a relationship that works for both the parent and the child in a meaningful and supportive way.

A Lifelong Journey

Building meaningful connections is not a destination, it's a lifelong journey.

Relationships evolve, deepen, or sometimes fade. The focus should not be on perfection but on authenticity and growth. For autistic individuals, this journey is deeply personal and is shaped by their unique strengths, needs, and perspectives.

In nurturing relationships, individuals with ASD enhance their own resilience and enrich the lives of those around them—as these connections grow, they remind us all of the power that human connection has to transform challenges into opportunities for joy and growth.

The lifelong journey of building relationships also emphasizes the importance of adaptability. Connections must evolve as circumstances change—whether through life transitions, new opportunities, or personal growth. Friendships may wax and wane, but the underlying commitment to building meaningful relationships remains a constant source of strength.

The Power of Human Connection

Relationships, with all their challenges and rewards, are the cornerstone of a fulfilling life. For autistic adults, these connections offer not only social interactions but a vital pathway to resilience and self-empowerment. By embracing the journey of connection—whether through family, friends, or shared passions—individuals with ASD can unlock the profound joy and support that relationships bring.

Romantic partnerships and parent-child relationships in particular require patience, understanding, and communication. A romantic partner can serve as a source of deep emotional support, but only when both individuals work toward mutual respect and clear communication. Love does not have to fit traditional molds to be meaningful—it can thrive in quiet gestures, shared routines, and thoughtful compromises. A relationship built on acceptance allows both partners to grow together, navigating challenges as a team rather than seeing differences as obstacles.

Similarly, parenting as an individual with ASD presents unique opportunities to redefine connections in ways that benefit both the parent and the child. Every child is unique, just as every parent is, and there's no single correct approach to building a loving, supportive relationship. The key is to embrace personal strengths while staying open to adaptation. By nurturing a sense of security, predictability, and love, parents with ASD can create an environment where their children thrive while they simultaneously honor their own needs. Relationships aren't about perfection, they're about connection. Whether as a partner, parent, or friend, the ability to build meaningful bonds enriches life in profound ways. For autistic individuals, these relationships can serve as a foundation for personal growth, self-acceptance, and a deep sense of belonging in a world that can sometimes feel overwhelming. Connection in *all* of its forms has the power to transform lives and bring warmth, understanding, and fulfillment.

Five Key Takeaways

Understanding Individual Social Needs

Autistic individuals experience relationships differently—they often require specific environments or communication styles to feel comfortable. Recognizing personal strengths, needs, and boundaries helps build authentic and sustainable connections. Whether through structured activities, written communications, or smaller gatherings, relationships should align with individual comfort levels.

The Importance of Clear and Direct Communication

Many social misunderstandings stem from differences in how people perceive emotions and intentions. Openly discussing feelings, expectations, and needs helps individuals with ASD navigate relationships more smoothly. Transparency fosters deeper connections in friendships, romantic partnerships, and parenting, and it also reduces misinterpretations.

Building Emotional Reciprocity and Understanding Social Cues

Expressing emotions in expected ways can be challenging for autistic individuals. While their emotional responses may not always align with societal norms, learning strategies such as mirroring emotions, providing verbal affirmations, and setting clear expectations can help them maintain healthy relationships. At the same time, neurotypical individuals should strive to understand and respect different ways of expressing connection.

The Role of Routine, Structure, and adaptability

Many individuals with ASD thrive on predictability. However, relationships require a degree of flexibility, too. Developing strategies such as structured planning, visual schedules, or prediscussed compromises helps maintain balance while respecting personal needs. Whether within the context of romantic relationships or parenting, finding a structure that works for both individuals strengthens their bonds.

The Power of Social Support and a Sense of belonging

Meaningful relationships provide emotional stability and resilience. Whether through close friendships, family, romantic partners, or support groups, creating spaces that embrace neurodiversity allows individuals with ASD to feel valued and accepted. Connection is not about conforming—it's about fostering mutual understanding, trust, and appreciation for different perspectives.

Self-Guided Activities to Build and Strengthen Relationships

Activity 1: Doing a Relationship-Mapping Exercise

Create a relationship map that identifies various connections in your life, such as family, friends, colleagues, and online communities. Note the strengths and challenges of each relationship, along with ways to improve communication and enhance understanding. This visual representation will clarify who your support networks are and will highlight areas where you may need to focus more on building connections.

Activity 2: Practicing Social Scripts

Practice daily social interactions by crafting social scripts for various scenarios, such as expressing gratitude, setting boundaries, or responding to a partner's feelings. Role-play these scripts with a trusted friend, therapist, or family member to build confidence in navigating real life interactions naturally and effectively.

Activity 3: Planning Joint Activities

Plan an activity with a romantic partner, child, or friend that aligns with your comfort levels and interests. Whether it's a structured event like a puzzle night, a shared hobby, or just having a simple conversation over coffee, this practice will promote mutual enjoyment while respecting individual preferences.

Activity 4: Writing in a Communication Reflection Journal

Maintain a journal to record conversations and social interactions and be sure to highlight your successes as well as your opportunities for improvement. Reflect on misunderstandings, emotional reactions, or difficulties you had with understanding the mental states of others. Over time, this practice will enhance your self-awareness and help reveal patterns in your social interactions.

Activity 5: Engaging in a Compassion and Understanding Challenge

Make a commitment to perform one act of compassion each day, whether it's directed at yourself or others. This could be as simple as recognizing your own progress, actively listening to a loved one's concerns, or sharing insights about neurodivergent perspectives with a friend. These kinds of small, intentional efforts will emphasize the value of empathy and foster stronger, more supportive relationships.

By actively engaging in these activities, you can strengthen your relationships with your loved ones in meaningful and sustainable ways. As we've already said, the goal is not to change who you are but to find ways to connect that feel authentic, fulfilling, and mutually supportive.

8

Embracing Mistakes: Transforming Challenges into Opportunities for Growth

Autistic individuals often feel overwhelmed. This can lead to increased stress, self-criticism, and a reluctance to embrace new experiences. The pressure to meet high standards— whether self-imposed pressure or pressure that is influenced by societal expectations—can result in individuals feeling intense frustration or even hesitating to try something again after having experienced a setback. However, avoiding mistakes altogether is neither realistic nor beneficial.

Mistakes are simply an unavoidable part of life.

Learning to see mistakes as valuable experiences encourages adaptability and self-acceptance. Every mistake offers an opportunity to refine skills, enhance problem-solving abilities, and boost self-confidence. By shifting the focus from self-judgment to self-compassion, individuals with ASD can foster a more positive relationship with learning and personal growth. Promoting a growth mindset—where effort and persistence are prioritized over immediate success—can be especially empowering. In supportive environments and with intentional reframing, mistakes can be transformed into powerful tools for growth rather than remaining sources of fear.

The Unique Challenges of Mistakes for Individuals with ASD

Interpreting emotions and collaborating with others involve various cognitive and behavioral processes that influence people's ability to engage socially. Mistakes with social interactions—misreading facial expressions, misunderstanding tone, responding in a way others don't expect—can feel discouraging, especially for individuals with ASD, and when such mistakes occur, they can lead to feelings of embarrassment or frustration. But recognizing different perspectives, adjusting to social cues, and gauging the right level of engagement are complex skills that require practice. When mistakes are made, those moments aren't signs of failure, they're opportunities for individuals to learn and refine their social understanding.

For many autistic adults, the fear of making social mistakes can lead to avoiding social situations altogether—they may hesitate to join conversations, ask questions, or express themselves due to past misunderstood experiences. However, avoiding social interactions can limit growth and increase feelings of pessimism and helplessness. In contrast, structured support can help people reframe mistakes as valuable feedback. Social skills training, role-playing, and guided practice can create safe spaces to explore interactions, learn from errors, and develop stronger communication skills.

Additionally, these supportive environments—whether in school, work, or community settings—play a critical role in reinforcing resilience. Encouraging patience, offering constructive feedback, and normalizing social missteps help individuals with ASD build confidence. This chapter explores the six fundamental abilities that are crucial for social development. (We've described them in past chapters, but we'll get into more detail here.) These six abilities illustrate how individuals with ASD can use mistakes as stepping stones to enhance their understanding, communication, and cooperative skills. By embracing social errors as part of the learning process, autistic adults can navigate interactions with greater confidence and success.

The Ability to Attribute Mental States to Oneself and Others

For people with ASD, understanding and interpreting mental states can be challenging. This ability—namely, Theory of Mind—is an essential aspect of social interactions and relationships. As we've already noted, "theory of mind" allows individuals to recognize that others may have different thoughts, feelings, and perspectives from their own. Many autistic individuals, however, struggle to grasp the idea that others' thoughts are separate from their own. For example, during a conversation, a person with ASD might assume that their knowledge is universally shared, leading to misunderstandings. This difficulty can occur in everyday situations, such as failing to recognize sarcasm or understand when someone is being deceptive. Developing ToM often requires structured learning experiences. Perspective-taking exercises can be particularly beneficial as they allow individuals with ASD to role-play different characters and consider how they might feel in various situations. Engaging in social stories can also be an effective strategy—these narratives illustrate common social scenarios and help autistic adults understand different perspectives. Encouraging individuals to ask themselves "What might this person be thinking?" fosters awareness of the mental states of others and promotes deeper social engagement.

Samantha, a young woman with ASD, struggled to understand why her friend was upset when she canceled plans at the last minute. Through guided discussions and social stories that her counselor used, Samantha learned to recognize that her friend had been looking forward to their time together and felt disappointed when their plans fell through. Over time, Samantha became more aware of how her actions impacted others. By continuously practicing these techniques, autistic individuals can enhance their ability to recognize and interpret the thoughts and emotions of others, leading to improved social interactions and deeper connections.

The Ability to Display Appropriate Emotional Reactions (Joint Attention of Emotion)

Joint attention of emotion involves sharing emotional experiences with others. This skill is essential for building empathy and strengthening social bonds. However, individuals with ASD often face challenges in responding appropriately to others' emotional states—due to differences in emotional processing, those with ASD may not always react in conventional ways to emotional cues. They might struggle to express sympathy when someone is distressed, for example, or have difficulty interpreting facial expressions and tone of voice.

To enhance their emotional reciprocity, autistic individuals can benefit from structured methods used by a therapist that are intended to foster emotional understanding. Teaching emotion labeling is highly effective—it enables individuals to recognize and articulate emotions in themselves and others. By also incorporating role-playing and modeling into daily routines, individuals can observe and practice appropriate emotional responses in various scenarios. Visual aids like emotion charts and guides for facial expressions further support emotional recognition, helping adults with ASD identify subtle cues that reveal another person's emotional state.

Josh, a high school senior with ASD, often seemed indifferent when his classmates were upset. Through structured learning and practice offered by a counselor, he began to recognize expressions of sadness. He learned to offer comforting words and gestures, which enhanced his social interactions. Over time, his ability to connect with his peers improved, and he developed stronger friendships. With continued exposure to emotional learning opportunities, autistic individuals can increase their capacity to interpret and respond appropriately to the emotions of others, resulting in enjoying more meaningful social relationships.

The Ability to Plan and Attend to Relevant Details in the Environment

Planning and attention to detail are essential for managing daily activities, solving problems, and setting goals. Many autistic individuals encounter challenges with executive function, which complicates their ability to plan and prioritize. Difficulties in planning can result in problems with managing time, completing tasks, and following multistep instructions. Individuals with ASD may feel overwhelmed when faced with unstructured tasks or changes in their routine.

To enhance their planning and organization skills, autistic individuals can utilize visual schedules, checklists, and planners to improve their ability to manage their time and responsibilities. Breaking tasks into smaller, manageable steps allows for more effective focus and execution. Setting alarms and reminders helps individuals stay on track and complete tasks efficiently. Establishing structured routines also provides predictability, reducing anxiety related to planning and organization.

Charlie, an adult with ASD, found it difficult to organize his work assignments. By using a task planner and setting specific deadlines, he was able to complete his work more efficiently and reduce stress. Over time, his ability to plan and prioritize tasks improved, making him more confident professionally and personally. Like Charlie, with continued practice and structured support, individuals with ASD can develop essential planning skills that will contribute to greater independence and success.

The Ability to Understand the Communicative Content of Gaze

Eye contact and gaze play a significant role in conveying emotions and intentions. Many autistic individuals find it challenging to interpret gaze-related cues, which impacts social interactions. Individuals with ASD may avoid eye contact or misinterpret the meanings behind the gazes of others, leading to social misunderstandings.

Understanding the communicative content of gaze requires structured learning experiences. Explicitly teaching gaze signals and their meanings can help autistic individuals recognize social cues more effectively. Video modeling—which is where individuals watch recorded social interactions with guided explanations—enhances comprehension of gaze behaviors. Practicing these scenarios in safe environments (such as during role-playing sessions with trusted individuals) allows for gradual improvement in recognizing and responding to gaze cues.

Gina, the college student with ASD who had difficulty looking at the eyes of others, also struggled to understand when someone wanted to speak to her. By practicing with a therapist who highlighted gaze cues, she became more skilled at recognizing when it was her turn to talk in conversations, and over time, she gained confidence in navigating social interactions. With consistent practice and support, individuals with ASD can better grasp gaze communication and enhance their ability to connect with others.

The Ability to Work Cooperatively with Others (Joint Attention of Behavior)

Collaboration and teamwork are crucial skills in both social and professional environments, yet individuals with ASD may find it challenging to coordinate their actions with others or grasp unspoken group dynamics. Many autistic people favor independent tasks, as group work can feel overwhelming due to unpredictable interactions and differing expectations. Structured social interactions and clear role assignments can foster cooperative behavior. Defining specific roles in group activities helps reduce ambiguity and provides a sense of control. Engaging in collaborative projects, such as board games and group tasks, allows individuals with ASD to practice working with others in a controlled setting. Encouraging turn-taking reinforces joint attention and helps build essential teamwork skills.

Ben, a 28-year-old with ASD, initially avoided team projects at work. With support from his supervisor, however, he learned to participate by taking on structured roles that played to his strengths (i.e., organizing data). This led to improved workplace integration. With continued guidance, he gained confidence in working cooperatively with others and ultimately strengthened his professional relationships.

The Ability to Understand, Comprehend, Analyze, Synthesize, Evaluate, and Differentiate Social Information in the Environment

Navigating social environments requires complex cognitive skills, from interpreting social cues to evaluating appropriate responses. Many autistic individuals experience difficulties in analyzing and responding to social dynamics—social situations often involve unwritten rules, making them difficult for individuals with ASD to decode. Misinterpreting humor, sarcasm, or indirect language, for example, can create barriers to meaningful interactions.

Providing structured opportunities to analyze social situations helps autistic individuals develop critical thinking skills, and engaging in discussions about various social scenarios and potential responses enhances social reasoning. Using social scripts can prepare individuals for real-life interactions by offering structured guidance, while pragmatic language training—focusing on tone, humor, and sarcasm—allows individuals to refine their communication skills.

An example of this is Ben, who often misinterpreted office humor, something that led to confusion. Through guided training in recognizing tone and intent, he gradually became more comfortable navigating workplace interactions. With continued practice, he strengthened his ability to effectively differentiate and respond to social information, allowing for more meaningful social engagement.

Understanding the Fear of Mistakes

Mistakes are a natural part of life, yet the fear of making them can feel overwhelming, especially for individuals who experience heightened anxiety or sensitivity to social judgments. This fear is particularly pronounced in autistic adults who may struggle with perfectionism, rigid thinking, or past experiences of being harshly judged for their mistakes. Many hesitate to try new things or engage in unfamiliar situations, fearing that any misstep will reflect poorly on them. The anxiety surrounding mistakes often arises from repeated experiences where errors led to criticism, rejection, or misunderstanding, reinforcing the belief that mistakes are unacceptable.

Recall Rebecca, a talented artist with ASD who struggled to present her work to others. She feared that imperfections in her art would reflect flaws in her ability, and that fear prevented her from sharing her talent. "If I show my paintings and someone points out a mistake, it feels like they're pointing out a mistake in me," she explained. The pressure to be perfect can create a paralyzing cycle, where avoiding mistakes leads to missed opportunities for learning and connection. This self-imposed pressure is not only about preventing embarrassment—it's about people protecting their sense of identity and self-worth.

For many adults with ASD, mistakes aren't simply errors to be corrected; rather, they can feel like fundamental failures that reinforce their insecurities about their competence and being accepted. This fear may extend beyond creative pursuits into academic, professional, and social settings. Some individuals might avoid speaking up in group discussions, trying new skills, or engaging in social interactions because they fear saying or doing something "wrong." Over time, this avoidance can result in stagnation, isolation, and a lack of confidence in their abilities.

Breaking this cycle requires a mindset shift—it's necessary to view mistakes not as personal shortcomings but as valuable opportunities for growth. Supportive friends, mentors, or therapists can provide significant encouragement, helping autistic individuals reframe mistakes as a normal and necessary part of learning. By gradually exposing themselves to situations where mistakes can occur and learning to tolerate discomfort, they can cultivate resilience and self-acceptance.

The Importance of Resilience in Navigating Mistakes

Resilience—the ability to cope effectively with adversity and setbacks—is essential when facing the fear of failure. Resilient individuals view mistakes not as endpoints but as opportunities for learning and redirection. For those with ASD, developing resilience involves recognizing personal triggers, setting realistic expectations, and nurturing a supportive mindset. Building this skill can take time, especially when past experiences have reinforced the belief that mistakes are unacceptable or result in negative consequences. However, we can all cultivate resilience through practice, reflection, and a willingness to adjust our perspective. Gerald, a middle school principal with ASD, exemplified this journey. After being promoted to a leadership role, he found himself second-guessing every decision—he feared the possibility of making mistakes that could impact his staff or students. The pressure to meet high standards weighed heavily on him, often leaving him feeling overwhelmed by the fear of making the "wrong" choice. Over time and with the input of a therapist, Gerald learned to view his role not as one that required perfection but as one that offered him the opportunity to model problem solving and adaptability. When he made a misstep during a staff meeting, he openly acknowledged it and used the instance to demonstrate how to recover and recalibrate. This transparency helped ease his anxiety and encouraged his entire team to tackle challenges with a growth mindset.

As Gerald discovered, resilience in leadership is not about avoiding mistakes, it's about demonstrating how to navigate them effectively. By reframing errors as learning experiences instead of failures, he fostered a culture of openness and adaptability within his school. His journey highlights the importance of self-compassion, constructive reflection, and the ability to separate self-worth from performance.

Learning from Mistakes: A Path to Growth

The journey to reframing mistakes starts with a shift in mindset. It involves moving away from self-critical narratives and recognizing that mistakes are not fixed judgments of ability but rather opportunities for growth. This shift is particularly significant for autistic individuals, who may face challenges with black-and-white thinking or have rigid views on success and failure. Learning to perceive mistakes as temporary setbacks, rather than definitive statements about one's worth or intelligence, can be transformative. However, this change doesn't occur overnight—it requires conscious effort, support, and consistent practice in real-life situations.

This mindset shift is closely related to attribution theory, which explores how individuals interpret the causes of their successes and failures. People who develop resilience tend to attribute their mistakes to external or temporary factors rather than viewing them as evidence of personal inadequacy. For instance, resilient individuals are more likely to think "I didn't have the right tools for this task" instead of "I'm bad at this." This distinction is crucial: attributing mistakes to changeable factors fosters adaptation and improvement, while viewing them as personal failings can lead to self-doubt and avoidance of challenges.

Lucy, a 26-year-old aspiring writer with ASD, struggled with getting rejections from literary publications. Initially, she perceived each rejection as proof that she lacked talent, reinforcing her belief that she wasn't "good enough" to succeed in writing. Every rejection letter felt like a personal indictment, making it difficult for her to continue to submit her work. However, with support from a writing group, Lucy gradually shifted her perspective on rejection—instead of viewing it as a total failure, she began to see it as part of the creative process. By analyzing feedback from editors and fellow writers, she learned to refine her craft. Over time, Lucy realized that rejection was not a reflection of her worth but an opportunity to improve. With this new perspective, she persisted, revising her work and resubmitting it.

Eventually, her persistence paid off when she published her first piece.

Lucy's experience highlights the power of reframing setbacks: she transformed rejection into motivation by shifting from a self-defeating mindset to one focused on growth. For autistic individuals, actively working to reinterpret mistakes—whether in creative, professional, or social settings—can lead to greater confidence and a more resilient approach.

Strategies for Embracing Mistakes

While shifting our perspective on mistakes is a deeply personal process, some strategies can help make this transition smoother and more effective. These approaches are significant for individuals with ASD, who often benefit from clear frameworks and actionable steps.

One key strategy is breaking down tasks into manageable steps. Rather than viewing a goal as a single, high-stakes event, breaking it into smaller, more achievable milestones can reduce the pressure of perfection. For example, if public speaking feels daunting, practicing in front of a mirror or with a trusted friend can provide a safe environment for growth before addressing a larger audience.

Another powerful approach is to focus on progress rather than outcomes. This involves celebrating minor improvements even if the overall goal has not (yet) been reached. Individuals can find this to be a transformative strategy when they're learning to navigate social situations— instead of fixating on whether a particular conversation went perfectly, they can begin to celebrate each time they initiated a dialogue or maintained eye contact regardless of the outcome.

Over time, these small victories can build confidence and comfort in social settings.

The Role of Self-Compassion in Learning from Mistakes

As we've highlighted in prior chapters, self-compassion plays a crucial role in overcoming the fear of failure. It involves treating ourselves with kindness and understanding, especially during difficult times. Unlike self-*criticism*—which magnifies mistakes and reinforces feelings of inadequacy—self-*compassion* fosters a balanced perspective that helps us acknowledge our struggles without allowing them to determine our worth. For autistic adults, cultivating self-compassion can be particularly challenging due to a history of being misunderstood, criticized, or subjected to unrealistic standards. Many have internalized the belief that they must meet high expectations to gain acceptance. This makes it more difficult to extend the same grace to themselves that they would offer to others.

Meg, a 30-year-old teacher with ASD, struggled with self-compassion after a lesson didn't go as planned. She initially blamed herself for not having prepared enough even though external factors like a fire drill had disrupted the class. The experience left her feeling frustrated and doubtful about her teaching abilities. However, with the assistance of a support group, she began to challenge these thoughts through mindfulness exercises and journaling. Rather than focusing solely on what went wrong, she learned to acknowledge the effort she had invested in her lesson and recognize that there had been circumstances beyond her control. "I realized that being kind to myself wasn't about ignoring mistakes, but about seeing them clearly and responding constructively," Meg reflected. This shift allowed her to move forward more confidently and adjust her teaching approach without being weighed down by self-recrimination. Compassion from others also plays a crucial role. When friends, colleagues, or family members respond to mistakes with patience and support rather than judgment, it reinforces the idea that errors do not define a person's worth. For individuals with ASD, having a trusted person to discuss challenges with can be transformative. Whether it's a mentor offering reassurance, a coworker helping to troubleshoot a problem, or a family member providing encouragement, external validation can strengthen the practice of self-compassion. Over time, this external support can assist individuals in developing their inner voice of kindness, making it easier to recover from setbacks.

That said, there is a caveat: building self-compassion should not be confused with ignoring responsibility. Rather, it's recognizing that imperfection is part of being human. When autistic adults learn to replace self-criticism with understanding, they become better equipped to handle mistakes, take healthy risks, and continue growing.

Reframing Mistakes as Valuable Information

Mistakes provide valuable information—they highlight areas for growth, reveal new approaches, and deepen self-awareness. When viewed through this lens, mistakes are feedback, not failures. This perspective is essential for individuals with ASD, who may struggle with rigid thinking patterns that make it difficult to see errors as opportunities rather than obstacles. Instead of interpreting mistakes as proof of incompetence, reframing them as valuable lessons can shift the focus from self-judgment to problem-solving. This shift reduces anxiety around failure and fosters resilience and adaptability.

Consider Anthony, a designer with ASD who faced repeated setbacks while designing a new prototype. Early in his career, he found mistakes deeply frustrating—he often felt like they reflected his limitations, not his progress. However, as he continued refining his designs, he began to view each failure as an opportunity rather than a roadblock. Each unsuccessful attempt provided critical insights into what hadn't worked and why, allowing him to make adjustments that brought him closer to his goal. "Every mistake was like a clue that led me to the bigger picture," he explained. His journey underscores that progress often arises from persistence, curiosity, and the willingness to learn.

Anthony's experience also highlights the role of experimentation and iteration in success. Many breakthroughs—whether in science, technology, or personal development—come from a series of refinements rather than instant perfection. By embracing mistakes as part of the learning process, autistic individuals can reduce their fear of failure and develop confidence in their ability to improve over time. This mindset benefits professional and academic pursuits and extends to social interactions, creative projects, and everyday problem-solving.

Building a New Relationship with Mistakes

Developing a healthier relationship with mistakes is a process that requires both inner reflection and practical action. It starts with confronting negative assumptions like "I'm not good enough" or "I'll never succeed." These thoughts, often automatic and deeply embedded, can create a self-fulfilling cycle of avoidance and self-doubt. Instead, individuals can practice cognitive reframing, i.e., substituting self-defeating thoughts with affirmations of their capability and potential. Statements like "I am learning through this experience" or "Mistakes help me improve" can shift the focus from self-criticism to self-growth. This mental shift doesn't mean ignoring mistakes but rather viewing them as constructive rather than defining.

It also involves setting realistic goals and allowing space for imperfection. Rather than striving for flawless execution, individuals can aim for meaningful effort. While perfectionism can paralyze people and make them unable to take risks or try new things, breaking tasks into smaller, more manageable steps can make challenges feel more approachable. Celebrating progress in small increments reinforces a sense of accomplishment and motivation.

This mindset fosters experimentation, creativity, and resilience. When individuals feel safe making mistakes, they're more inclined to explore innovative solutions, engage in problem solving, and develop new skills. In professional contexts, this can promote greater adaptability and confidence; in personal life, it encourages a sense of self-acceptance and growth. It's true that cultivating a healthier relationship with mistakes takes time, but with conscious effort and in supportive environments, this new relationship can become a powerful tool for lifelong learning and success. Whether those supportive environments are formed through therapy, mentorship, or peer groups, having a safe space to confront our fears and celebrate our progress can enhance our confidence and ability to adapt.

The Lifelong Journey of Growth

Mistakes aren't barriers, they're bridges. They connect us to deeper understanding, richer experiences, and greater resilience. For individuals with ASD, embracing mistakes as part of the human experience opens doors to personal and professional growth. It shifts the focus from a fear of failure to a mindset of curiosity and adaptability, fostering a sense of self-trust and confidence. Each mistake becomes an opportunity for autistic adults to learn, refine their skills, and strengthen their problem-solving abilities.

As this journey unfolds, it's important to remember that growth is not linear. There will be moments of doubt, frustration, and challenge. Some setbacks may feel discouraging, and progress may come in small, gradual steps rather than giant leaps. Yet, each step forward—no matter how small—is a testament to courage and the capacity for transformation. Recognizing the value of persistence helps individuals with ASD stay motivated even when the path ahead feels uncertain.

Five Key Takeaways

Mistakes Are Opportunities for Growth

Mistakes are an unavoidable part of life; avoiding them is neither realistic nor beneficial. For autistic individuals, mistakes can feel overwhelming and lead to stress and self-criticism. However, reframing mistakes as valuable learning experiences fosters adaptability, resilience, and self-acceptance. Every mistake provides a chance for adults with ASD to refine their skills, enhance their problem-solving abilities, and build their confidence.

The Fear of Mistakes Can Lead to Avoidance

Many individuals with ASD struggle with the fear of making mistakes, especially in social interactions. Past experiences of judgment or criticism can make them hesitant to try new things or engage with others. While this avoidance is understandable, it limits opportunities for learning and personal growth. Overcoming this fear requires structured support, encouragement, and a shift toward seeing mistakes as usual and essential for development.

Resilience Helps Individuals Navigate Mistakes

Resilience—the ability to cope with setbacks—is crucial in overcoming the fear of mistakes. Autistic individuals may struggle with rigid thinking and perfectionism, traits that make it difficult to handle errors, but building resilience through practice, reflection, and support helps them view mistakes as stepping stones rather than failures. Embracing mistakes as part of the learning process will foster confidence and their ability to adapt to their situations and circumstances.

Self-Compassion Reduces the Negative Impact of Mistakes

Self-criticism magnifies mistakes, leading to feelings of inadequacy and discouragement. Instead of harshly judging themselves, individuals with ASD can benefit from self-*compassion*, which involves acknowledging their efforts and treating themselves with kindness. Recognizing that everyone makes mistakes and learning from them rather than dwelling on them creates a healthier approach to personal and professional growth.

Reframing Mistakes as Valuable Information Encourages Growth

Instead of viewing mistakes as personal shortcomings, individuals can see them as feedback that highlights areas of possible improvement. This perspective shift reduces anxiety and promotes problem-solving skills. From scientists to artists, many successful individuals have used mistakes as a tool to refine their work and their processes. By embracing mistakes as part of the learning experience, individuals can take healthy risks, develop new skills, and confidently grow.

Self-Guided Activities to Manage Mistakes

Activity 1: Reframing Negative Thoughts Exercise

Practice identifying negative thoughts about mistakes and reframe them more positively. For example, instead of thinking, "I failed, so I'm not good at this," you can say, "This didn't work out, but I can try a different approach next time." By writing down these thoughts in a journal, you'll reinforce a healthier mindset over time.

Activity 2: Role-Playing Social Scenarios

You may struggle with social mistakes, such as misreading cues or responding in unexpected ways, but role-playing social interactions with a trusted friend, therapist, or coach can give you a safe space where you can make and learn from mistakes. Practicing different responses in controlled settings will allow you to build up your confidence to be able to better navigate real life interactions.

Activity 3: Breaking Tasks into Manageable Steps

Break a goal into smaller, more achievable steps. For instance, if public speaking intimidates you, you can start by practicing in front of a mirror and then move on to speaking in front of a small group. Large tasks can feel overwhelming, which increases the fear of making mistakes. In this activity, you'll reduce your stress and focusing on progress rather than perfection.

Activity 4: Self-Compassion Meditation or Journaling

Self-compassion exercises like guided meditation or reflective journaling can help you process mistakes without excessive self-judgment. When you write about a mistake, what you learned, and how you would support a friend in a similar situation, that will encourage you to be kinder to yourself and will reduce your fear of failure.

Activity 5: Creating a "Mistakes and Lessons Learned" Log

Keeping a personal log of mistakes and the lessons they provide can help shift your perception of failure. Instead of viewing errors as being unfavorable, you can reflect on how each mistake contributed to your personal growth. Over time, this record will reinforce the truth that mistakes are part of progress, not setbacks.

By engaging in these exercises, you will be less afraid to make mistakes and more confident to face new challenges. An ongoing fear of failure can lead you to avoid opportunities that have the potential to strengthen success and satisfaction in both your personal and professional lives.

9

Embracing Success by Focusing on Strengths

Recognizing and celebrating personal achievements is crucial in building confidence and resilience; acknowledging even the most minor successes can significantly boost one's sense of accomplishment and help pursue personal growth. Everyone has unique strengths, which we have called "islands of competence"—areas where natural talent or ability shines. Individuals gain greater self-assurance by identifying and developing these strengths; this mindset will allow them to navigate obstacles more confidently.

Creating an environment that supports and nurtures these strengths further enhances personal development. Individuals who take pride in their accomplishments are more likely to view their successes as small steps toward future progress. This perspective fosters a positive self-image and a proactive mindset. For individuals with ASD, recognizing these "islands of competence" is particularly important, as it provides opportunities to experience a sense of mastery and capability. These strengths can serve as a foundation for greater self-confidence, allowing individuals to engage more fully in various aspects of life and overcome challenges with renewed determination.

Embracing Success Through Developing Cognitive Strengths

Success isn't just about overcoming every challenge; it also involves finding ways to utilize personal strengths to navigate difficulties. For autistic adults, understanding and adapting to social and cognitive challenges is crucial for both professional and personal growth. By concentrating on structured strategies, leveraging logical thinking, and exploring alternative approaches to social interactions, individuals can build confidence and achieve success in ways that align with their natural abilities. The following sections will explore specific cognitive traits that individuals with ASD often struggle with and how they can harness their strengths to transform challenges into opportunities for growth.

The Ability to Attribute Mental States to Oneself and Others

As we've emphasized, understanding the perspectives of others is vital for achieving personal and professional success. Although many autistic adults excel in logical analysis, they often struggle to interpret emotions and intentions. Take Mark, a data analyst who shines in problem-solving yet frequently misreads his colleagues' silence as indifference. He assumes that if a coworker is quiet in a meeting, they're disengaged; he doesn't consider that they might be deep in thought or processing information.

This challenge arises from difficulties in attributing mental states (beliefs, desires, emotions) to oneself and others. Recognizing that different people may interpret the same situation in various ways is a fundamental aspect of social cognition. If someone lacks this skill, social interactions can lead to them feeling confused or anxious. Mark has experienced this firsthand in team collaborations, where subtle social cues often shape the course of conversation. Fortunately, Mark has developed strategies to bridge this gap. Seeing as his strength lies in his analytical abilities, he has learned to supplement his social understanding with structured techniques. He maintains a mental checklist of possible emotional responses and reflects on past interactions to enhance his interpretations, and he also actively seeks feedback from trusted colleagues to ensure that his assumptions align with reality. Additionally, he employs direct questions to clarify intent, such as asking, "Would you like input, or do you need a moment?".

Mark's approach has helped him improve his workplace interactions while leveraging his existing strengths. Over time, he has learned to recognize patterns in human behavior, which allows him to adjust his responses accordingly. By combining his analytical skills with deliberate social strategies, Mark has developed effective ways to navigate complex interpersonal dynamics.

The Ability to Display Emotional Reactions Appropriate to Another Person's Mental State (Joint Attention of Emotion)

Success in relationships and the workplace often depends on responding appropriately to the emotions of others. The ability to engage in joint attention of emotion—recognizing and mirroring another person's emotional state—fosters connection and trust. However, for some individuals with ASD, understanding and displaying expected emotional reactions in social interactions can be challenging.

Jane is an engineer who excels at designing complex systems but struggles with emotional reciprocity. When a coworker expresses excitement about a promotion, Jane instinctively responds with a factual analysis, comparing salary differences or discussing industry trends instead of expressing enthusiasm. While her insights may be valuable, they can come across as detached or unfeeling, leaving her coworker feeling unheard.

Jane's strengths lie in her attention to detail and her logical reasoning. However, she has recognized the importance of emotional expression in social settings and has worked to develop strategies to enhance her interactions. One technique she has adopted is incorporating verbal affirmations before engaging in analysis. For example, she now consciously tries to say, "That's great news!" or "Congratulations!" before discussing the implications of the promotion. Additionally, Jane has learned to observe her colleagues' emotional cues more closely. She practices pausing before responding, giving herself time to assess whether the situation calls for empathy, celebration, or problem-solving. Over time, these small but deliberate changes have made a significant difference in her relationships, helping her build stronger connections while remaining true to her natural way of thinking.

By balancing her analytical mindset with intentional social strategies, Jane has improved her ability to engage in joint attention of emotion. These adjustments strengthen her workplace interactions and contribute to a more supportive and collaborative environment.

The Ability to Plan and Attend to Relevant Details in the Environment

Planning and attention to detail are crucial for professional success, but autistic individuals may struggle with prioritization. While a strong focus on detail can lead to high-quality work, it may also result in inefficiencies if lower-priority tasks receive too much attention.

Russell, a software developer who excels at coding, often becomes fixated on optimizing minor features instead of completing high-priority tasks. While his meticulous approach ensures that his code is clean and efficient, it has also resulted in missed deadlines and unnecessary stress. His managers have noted that while his work is technically outstanding, his tendency to overfocus on trivial optimizations sometimes delays crucial project milestones.

Recognizing this challenge, Russell leveraged his analytical skills to create a structured prioritization system. He developed a checklist that categorizes tasks by impact and deadline, allowing him to visually assess what requires immediate attention. By incorporating a weighted scoring system—assigning values to urgency, importance, and effort—he can objectively determine the execution order.

Additionally, Russell has adopted time-blocking techniques to set aside specific periods for optimization—that way, it doesn't interfere with his ability to meet project deadlines. He also employs accountability strategies, such as checking in with a mentor or project manager to confirm that his focus is aligning with team priorities.

By applying his logical strengths to improve his workflow efficiency, Russell has enhanced his ability to manage tasks without changing his fundamental approach to problem solving. His structured system allows him to balance quality with efficiency, ultimately reducing stress and improving project outcomes for everyone.

The Ability to Understand the Communicative Content of Gaze

Nonverbal communication is key to professional and personal success, often conveying as much meaning as spoken words. Eye contact is a particularly powerful tool in social interactions as it signals attention, interest, or shifts in conversation. However, for individuals with ASD, interpreting eye gaze can be challenging. They may struggle to discern when someone is seeking engagement, offering a cue for turn-taking, or signaling a change in topic. Linda, for example, is a successful accountant who excels at financial analysis but found it challenging to interpret eye contact. During meetings, she often missed subtle cues that indicated when someone wanted to speak, needed clarification, or was ready to transition to a new subject. While her expertise and analytical skills made her an invaluable asset to her team, these nonverbal challenges sometimes led to miscommunications or moments of awkward silence.

Instead of forcing herself to maintain eye contact in a way that felt unnatural and overwhelming, Linda developed an alternative strategy: she focused on other visual cues like body positioning, facial expressions, and hand gestures to understand the flow of conversation. By paying attention to when a colleague leaned forward, raised an eyebrow, or shifted their posture, she was able to accurately gauge engagement and participation without having to rely solely on making direct eye contact.

Additionally, Linda has incorporated structured techniques into her communication style. For instance, she set clear expectations at the beginnings of meetings and asked colleagues to use verbal signals like "I have a question" or "Let's move on to the next point." This ensured that discussions remained smooth and inclusive.

By leveraging her observational skills and developing personalized strategies, Linda had devised a strategy to successfully navigate social interactions while staying true to her strengths. In short, rather than forcing herself to conform to traditional norms of eye contact, she found ways to engage effectively and confidently in professional settings.

The Ability to Work Cooperatively with Others (Joint Attention of Behavior)

Teamwork is essential to many professions; it requires individuals to coordinate, communicate, and adapt to different working styles. However, collaborative environments can sometimes feel overwhelming for autistic individuals due to unstructured group dynamics, unclear expectations, and the need for spontaneous social interactions. While many professionals thrive in team settings, others—like Chris, a mechanical engineer—find group work challenging despite exceling in independent projects.

Chris preferred structured, clear instructions and struggled when expectations were ambiguous. In group settings, differing communication styles, vague task assignments, and shifting priorities created frustration and confusion for him. But rather than disengaging or feeling overwhelmed, Chris took a proactive approach to redefine his role in collaborative projects: He became the team organizer. Now he uses his detail-oriented mindset to create clear task lists, assign responsibilities, and set deadlines, ensuring that each member knows their role. By implementing structured workflows, he minimized confusion and helped the entire team collaborate more efficiently.

Additionally, Chris introduced regular check-in meetings to keep projects on track and provide a structured space for feedback and adjustments. His method helped his colleagues stay organized and allowed him to contribute in a way that aligned with his strengths. Through this approach, he has transformed his teamwork experience from being a source of stress into becoming a structured and productive process.

The Ability to Understand, Comprehend, Analyze, Synthesize, Evaluate, and Differentiate Social Information

Although social complexity can be challenging, autistic individuals often excel in pattern recognition and structured analysis. Still, navigating interpersonal relationships in professional settings requires interpreting subtle cues, managing expectations, and adapting to shifting dynamics. Unspoken rules and social ambiguity can be overwhelming in this area.

Maria is a skilled marketing strategist. She exceled at analyzing consumer behavior and data trends, but she struggled with office politics—she found social nuances confusing and often missed the unspoken expectations of workplace relationships. Casual small talk, implicit hierarchy rules, and shifting alliances among colleagues made workplace interactions unpredictable and stressful for her.

Given her difficulty with intuitively deciphering unstructured social norms, Maria decided to apply her analytical mindset to social interactions. She treated workplace dynamics like a system: she observed communication patterns, identified trends in her colleagues' behavior, and adjusted her approach accordingly. For instance, she recognized which coworkers preferred direct communication and which ones appreciated a more relationship-driven approach.

To refine her strategy further, Maria kept a mental or written record of past interactions, noting which responses were effective and which had led to misunderstandings. Using logic and pattern recognition, she structured her approach to networking, meetings, and collaborations, ensuring that she could engage effectively while minimizing uncertainty. Additionally, Maria learned to ask clarifying questions to uncover hidden expectations. Instead of assuming she understood a vague directive, she sought specificity by asking, "Would you like me to focus on the data, or would you prefer a broader market analysis?" This method allowed her to navigate professional relationships more confidently while maintaining her strengths in structured thinking.

Through this approach, Maria has successfully bridged the gap between social ambiguity and structured analysis. This allowed her to navigate workplace relationships while remaining authentic to her natural skills.

How We Think About Success

Success isn't just about reaching a goal, it also includes how we view and reflect on our achievements. Some people take pride in their hard work and effort when they succeed. They often say, "I worked hard and am ready for more challenges." Others, including many with ASD, don't feel the same way—they think, "I was lucky. I don't believe I can do that again."

Understanding these different perspectives on success is essential because when we believe that our achievements stem from our effort and skills, we feel more confident in facing challenges. However, if we think our success is merely due to luck, it's hard to feel secure about trying again. Success based on confidence and effort establishes a solid foundation for growth and resilience. This perspective also helps us handle mistakes and setbacks because we trust in our ability to recover and move forward.

The Role of Satisfaction

Feeling satisfied with what we achieve is a key part of building resilience. Satisfaction provides a sense of accomplishment, reinforcing positive behaviors and encouraging further growth. When individuals who display confidence succeed, they feel joy and take a moment to celebrate. They acknowledge the effort they've put in—they recognize that their hard work has led to positive outcomes. This ability to internalize success strengthens their motivation, pushing them to pursue new challenges confidently.

However, not everyone finds it easy to embrace success. Some people struggle to feel satisfied even after they've achieved their goals. Instead of celebrating their accomplishments, they might dismiss them as mere luck or feel like they don't deserve recognition. This mindset— which is often associated with imposter syndrome—can prevent people from fully appreciating their progress and can lead to ongoing self-doubt.

Recognizing and taking pride in accomplishments, no matter how small, is crucial for long-term motivation and well-being. One effective strategy is for individuals to keep a success journal where they can write down their achievements and note the effort they put into them.

Reviewing these notes can serve as a reminder of their capabilities and progress.

Additionally, sharing accomplishments with a supportive network—whether that consists of friends, mentors, or colleagues—can help reinforce a sense of achievement. When others acknowledge and celebrate our successes, it becomes easier to internalize them.

Feeling satisfied with our progress doesn't mean being complacent, though! Instead, it fosters a positive cycle of motivation. Allowing ourselves to take pride in our accomplishments boosts our confidence, encourages us to be persistent, and makes it easier for us to embrace new challenges with enthusiasm.

Being Proud Without Boasting

Someone taking credit for their achievements isn't the same as bragging. True success is built on confidence, not the need for external validation, and when people feel comfortable about their accomplishments, they don't need to announce them to the world to feel valued. Genuine pride in success is quiet and secure, rooted in self-awareness rather than comparison. It's about recognizing our own hard work and abilities without diminishing the efforts of others. Instead of boasting, we let our actions and continued success speak for themselves. For example, a top performing athlete doesn't need to constantly remind others of their victories—their dedication and results make their achievements evident.

Leaders who are humble about their success are often more respected and effective in their roles. They acknowledge their efforts while also appreciating the contributions of mentors, teammates, and colleagues. This humility strengthens relationships, fosters trust, and creates an environment where others feel valued and motivated to succeed. Success isn't about chasing admiration, but rather appreciating personal growth and continuous improvement. Those who adopt this mindset can maintain long-term motivation and inspire others through their commitment and integrity.

The Fear of Success

Sometimes people feel anxious about or afraid of succeeding. This fear may stem from self-doubt, a belief that they don't deserve success, or concerns that greater achievements will lead to increased pressures and expectations. The fear of success can manifest as a hesitation to take on new challenges, avoidance of opportunities, or self-sabotaging behaviors that prevent individuals from reaching their full potential.

Carla, a young woman with ASD, felt nervous when she was offered a promotion at work. She was initially excited but soon began doubting herself, wondering, "What if I can't handle the new responsibilities?" and "What if people expect too much from me?" These thoughts created stress and made her question whether she should accept the promotion. Autistic individuals often encounter challenges with transitions and uncertainty, which can make professional growth feel overwhelming. However, when Carla expressed her concerns to a mentor, she realized that she did indeed have the skills and support necessary to succeed. She also came to understand that fear is a natural response to change and didn't indicate any kind of inability on her part.

By focusing on her strengths—namely, her attention to detail, her problem-solving skills, and her ability to persevere—Carla gradually grew more confident in her ability to take on new challenges. She also developed coping strategies, like breaking tasks into smaller steps and seeking advice from trusted colleagues. With this shift in mindset, she accepted the promotion and realized that she could manage her new role well.

Overcoming the fear of success involves recognizing our self-worth, acknowledging our abilities, and embracing challenges as opportunities for growth. By shifting our focus from fear to potential positives, we can step into success with confidence and resilience.

Finding Islands of Competence

Everyone possesses unique strengths—those are each person's islands of competence. They signify areas where we naturally excel, whether it be a skill, a hobby, or an activity that brings us joy and fulfillment. By recognizing and building on these strengths, we establish a foundation of confidence that assists us in overcoming challenges in other areas of life.

For example, Tim discovered that woodworking was one of his areas of competence— crafting furniture gave him joy and a deep sense of accomplishment, and each completed project reinforced his belief in his abilities and reminded him that he could succeed. Over time, the confidence he gained from woodworking spilled over into other aspects of his life. He became more willing to take on challenges at work, speak up in social settings, and pursue new learning opportunities.

A person's islands of competence do more than showcase strengths— they also reveal what brings that particular person happiness and fulfillment. For some, this may include artistic expressions like painting or music; for others, their islands of competence might involve problem-solving, organizing, mentoring, or even performing acts of kindness. Acknowledging these strengths allows people to cultivate them intentionally and use them as sources of motivation and self-confidence.

When people focus on their strengths, they also build up their resilience. Getting involved in their areas of expertise can provide reassurance and stability when difficulties pop up. If someone struggles with social anxiety but excels at writing, for example, they might use written communications to connect with others. Essentially, when individuals recognize their valuable skills, they gain a greater sense of self-worth and challenges become easier to manage.

Islands of competence are not just about success but about self-discovery and growth. By embracing what they do well, individuals can create a stronger, more confident version of themselves and be prepared to take on new challenges with a positive mindset.

Making Changes Step by Step

The first step in embracing success is to reflect on how we perceive it. Do we acknowledge our achievements, or do we dismiss them as simply being luck or due to external factors? Many of us struggle with self-doubt and think we haven't genuinely earned our successes. This mindset can prevent us from fully appreciating our progress and recognizing our abilities. But by evaluating our beliefs about success, we can begin to make positive changes in how we see ourselves and our accomplishments.

The next step is to challenge our negative thoughts. If we often think "I don't deserve this" or "I was just lucky," we can try replacing those thoughts with more constructive and realistic alternatives. Instead of attributing success to luck, we can recognize that our hard work, skills, and dedication played a significant role. A helpful strategy is to keep a journal to document achievements and the effort that was involved in attaining those achievements—reviewing this journal can reinforce the fact that our success is well-earned. Another practical approach is to practice self-affirmations, such as saying aloud "I worked hard for this" or "I am capable and deserving of success." Over time, these intentional mindset shifts can help build greater confidence.

The final step in this process is to concentrate on our interests and strengths. We should participate in activities that bring us joy and showcase our skills. Whether it involves problem solving, creative expression, teaching, or a particular hobby, engaging in what we excel at reinforces our sense of self-worth. These activities act as a grounding force, especially during challenging times—that's when it's especially crucial that we're reminded of our competence and value.

We've emphasized throughout this book that making changes does not happen overnight. Rather, the process takes place gradually as we reflect on our mindset, challenge negative thoughts, and nurture our strengths. These activities can build confidence and a healthier relationship with success.

Celebrating Successes: An ASD Perspective

Success isn't solely defined by major accomplishments—success also includes the smaller victories that foster a sense of pride and fulfillment. Even though society often equates success with career milestones, financial accomplishments, or broad recognition, for autistic individuals, success may manifest in mastering a new routine, navigating social interactions, adapting to changes, or finding comfort in personal interests. Success might involve developing new coping strategies, creating a sensory-friendly environment, or effectively expressing needs. It can be as straightforward as getting through an overwhelming day, trying new foods, or engaging in conversation. These daily achievements deserve as much recognition as more traditional successes do. Acknowledging and celebrating these achievements regardless of their magnitude can enhance confidence and instill a sense of progress.

As Ralph Waldo Emerson said, "To know that even one life has breathed easier because you have lived—this is to have succeeded." Success isn't solely about personal milestones! It also involves making a difference, whether through sharing knowledge, offering kindness, or simply being authentic. Small acts of self-advocacy, helping others understand autism, and forming meaningful connections all contribute to making positive impacts.

Success is deeply personal and must be defined by individual strengths and aspirations rather than by external expectations. Growth may appear different for each person, but nonetheless, every step forward holds meaning. Celebrating progress, no matter how small, encourages self-acceptance, motivation, and resilience. Taking time to recognize achievements like managing sensory overload, maintaining focus, or embracing a passion builds confidence and establishes a foundation for future growth.

In a world that often prioritizes neurotypical standards of success, it's essential to recognize and embrace the unique victories that come with being autistic. Every achievement, no matter how personal, holds value. By focusing on strengths, acknowledging progress, and celebrating small wins, autistic individuals can cultivate a fulfilling and authentic sense of success.

Five Key Takeaways

Emphasizing Strengths Builds Confidence

Success for individuals with ASD is best achieved by recognizing and focusing on their strengths, also known as "islands of competence." These strengths provide a foundation for building self-esteem and navigating challenges with confidence. Instead of dwelling on difficulties, individuals can leverage their natural abilities to create fulfilling experiences.

Structured Approaches Lead to Success

Many autistic individuals thrive when they use structured strategies to navigate social and professional environments. By developing routines, checklists, and logical approaches, they can turn potential obstacles into opportunities for success. These strategies allow them to function in ways that align with their cognitive strengths.

Understanding Social Cues Can Be Learned

While social interactions can be challenging, individuals with ASD can develop methods to interpret emotions and body language using structured techniques. For instance, employing direct questions or paying attention to physical cues instead of concentrating on eye contact can improve communication skills. These adaptive strategies make social engagement more accessible and comfortable.

Confidence in Achievements Shapes Success

The way individuals interpret their achievements dramatically impacts their future motivations. Those who attribute success to their effort and skills are more likely to take on new challenges, whereas those who believe success is due to sheer luck may feel hesitant. Developing a growth mindset helps individuals embrace future opportunities with confidence.

Celebrating Small Successes Enhances Resilience

Success isn't just about reaching major milestones—it's also about recognizing and celebrating small victories. Taking pride in accomplishments, no matter how minor, fosters a sense of progress and resilience. By acknowledging each step forward, autistic individuals can develop a more positive outlook on their personal and professional growth.

Self-Guided Activities to Enhance Success

Activity 1: Doing a Strengths-Mapping Exercise

Create a list of your strengths, a.k.a. your "islands of competence." Jot down activities you enjoy and tasks you excel at. Determine how to integrate these strengths into your daily life, whether at work, in relationships, or through hobbies. This exercise will enhance your self-awareness and confidence by encouraging you to emphasize what makes you unique.

Activity 2: Practicing Social Strategies

Choose a common social scenario like a work meeting or a casual conversation. Plan different ways to respond using structured strategies, such as asking direct questions or observing body language. Practicing these techniques with a trusted friend or writing about them in a journal can help you refine your social skills in a natural way.

Activity 3: Developing Tools for Prioritization and Task Planning

Develop a task management system suited to your strengths. This could include color-coded lists, time-blocking strategies, or digital reminders. Organize tasks based on urgency and importance to boost your efficiency in your work and with your daily responsibilities. This practice supports productivity by helping you avoid getting stuck on minor details.

Activity 4: Reframing Success by Journaling

Maintain a journal to document your daily successes, no matter how minor they might seem. After each entry, note what contributed to that success—whether it was effort, skill, or preparation. This practice will help you strengthen the belief that success stems from personal ability rather than luck, a mindset that will give you greater confidence and resilience over time.

Activity 5: Creating a System for Celebrating Rewards

Create a personal reward system to acknowledge your achievements. Set goals and select meaningful rewards, such as treating yourself to a favorite activity after finishing a challenging task. By purposefully celebrating your progress, you'll train your mind to recognize and appreciate your successes, thereby cultivating a positive cycle of motivation.

These key takeaways and exercises will provide you with practical strategies for embracing success in ways that align with your strengths and natural abilities.

10

Self-Discipline and Control: Strategies for Personal Growth

Self-discipline and self-control are essential skills that act as building blocks for personal growth, independence, and living a fulfilling life. These abilities enable us to manage our time effectively, maintain focus on important tasks, regulate our emotions, and make thoughtful decisions. For individuals with ASD, developing self-discipline can offer stability, a sense of control, and the confidence they need to navigate both everyday challenges and long-term aspirations.

In this chapter, we'll explore practical strategies for strengthening self-discipline and self-control, and we'll also provide tools and insights to support personal development. Remember that self-discipline is *not* about achieving perfection! Instead, it involves making small, consistent changes that help us feel more capable, independent, and resilient. Each step forward builds momentum, making it easier to reach our goals and lead a more balanced and fulfilling life.

The Meaning of Self-Discipline

Self-discipline includes making decisions and taking actions that align with our values and goals even when we're faced with distractions or challenges. It involves resisting the urge to act impulsively, staying calm when emotions run high, and finding ways to keep progressing even when situations feel tough. For instance, completing a task before watching TV or responding thoughtfully in a tense situation both demonstrate self-discipline.

Developing self-discipline involves learning to prioritize long-term rewards over short-term gratification, focusing on building habits that promote consistency—for example, following a study routine, adhering to a budget, or committing to a fitness plan. Individuals who exhibit strong self-discipline often employ techniques like setting clear goals, breaking tasks into manageable steps, and utilizing reminders or accountability systems to stay organized. Rather than being about maintaining rigid control, self-discipline is about making conscious choices that align with our greater purpose.

Self-control is closely tied to self-discipline. It involves managing our responses to situations and emotions. It provides a chance to pause and reflect before acting, actions that help prevent misunderstandings, mistakes, and conflicts. Practicing self-control enables us to face challenges with clarity; it also reduces impulsive reactions and promotes thoughtful communication.

Self-discipline and self-control form a strong foundation for personal growth and emotional balance. Cultivating self-discipline is a gradual process. It requires practice, patience, and a willingness to learn from experience. Over time, these qualities help build resilience, sharpen focus, and reinforce a sense of inner stability. Each small effort contributes to individuals being able to live in a way that makes them feel empowered and prepared for challenges and opportunities alike.

Self-Discipline and Control Challenges for Autistic Individuals

Executive Function Challenges

Executive function skills form the foundation of self-discipline; those skills encompass the ability to plan, prioritize, initiate, and complete tasks. Adults with ASD often face significant challenges in this area, which directly impacts their capacity for self-regulation and consistent behavior. For instance, cleaning an apartment may seem simple, but it actually requires several cognitive steps: determining where to start, assessing necessary supplies, managing energy levels, and maintaining focus. Many autistic individuals may find themselves immobilized during the decision-making process or feel overwhelmed by the numerous steps involved, leading them to abandon the task despite their initial motivations.

Seth is a middle-aged man with ASD who was deeply passionate about building intricate model trains. This hobby provided him with comfort and focus. However, his intense interest often led to him to hyper-focus on trains and consumed many hours of his free time; his hobby interfered with essential tasks like responding to work emails and preparing meals. With a structured visual schedule and external time cues, he learned to allocate specific hours to his favorite activity *and* preserve some energy for his daily responsibilities. By structuring transitions between tasks, he was able to enjoy his interests without sacrificing any of his work commitments.

This gap between intention and action contributes to frustration and negative self-perceptions, like someone thinking, "Why can't I do something so basic?" That experience further hinders future attempts at self-discipline. But these challenges don't stem from laziness or a lack of willpower—they arise from neurocognitive differences. External support is essential to bridge this gap. Tools such as visual planners, color-coded schedules, digital reminders, and task breakdown sheets act as extensions of executive function, aiding the mental processes that individuals need to follow through on their tasks. Consistently using such aids helps individuals strengthen their self-discipline skills, not by changing their neurology but by working with it.

These external supports reduce cognitive friction and make it much easier to complete tasks.

Emotional Regulation and Control and Social-Cognitive Differences

Emotional self-control is a crucial element of self-discipline. Its importance becomes even more evident for autistic adults, who often experience heightened emotional sensitivity. Stressors that may only mildly irritate others—loud noises, unexpected schedule changes, even a missed message—can trigger overwhelming emotional reactions in individuals on the spectrum. This emotional reactivity can disrupt executive function, making it challenging to prioritize tasks, maintain focus, or engage in rational problem-solving in the moment. Self-discipline does not falter due to a lack of motivation—it falters because the body and brain enter a state of dysregulation.

For instance, receiving constructive criticism can disrupt an entire afternoon. The emotional impact may linger for hours or even days, making it challenging to remain productive or maintain routines. This isn't an overreaction—it reflects a distinct neurological pattern that amplifies emotional responses. Consider Gina, the college student we discussed in Chapter Seven. She found it challenging to interpret the tone of group messages, perceiving a short or ambiguous reply from a friend as rejection. This led Gina to become anxious or socially withdraw even when no harm was intended.

Another example is April, a young woman with ASD who struggled to interpret her boyfriend's facial expressions or grasp certain gestures that carried emotional weight, such as unexpected hugs. Her confusion often resulted in her withdrawing or miscommunicating, even when she meant to convey affection.

There are strategies to help navigate these challenges. Such tactics include keeping a structured journal to track emotional patterns, setting alerts for regular breaks to self-regulate, or engaging in guided movement practices like yoga or tai chi—all of those activities can promote better emotional balance and greater emotional resilience. Other examples are scripting responses, employing visual cues for emotional states, practicing mindfulness techniques, and creating sensory-safe spaces.

Developing self-awareness and having regulatory tools readily available supports consistent behavior even when emotions peak. Recognizing early signs of dysregulation and building routines around recovery can provide a powerful sense of stability for times when executive function is under pressure. Over the long term, fostering emotional literacy and practicing de-escalation techniques enables autistic adults to stay grounded during moments of both calm and distress.

Contextual Challenges in Applying Self-Discipline

Self-discipline is often regarded as a universal trait—once mastered, it can be applied in any situation. However, for adults with ASD, a lesser-known challenge in executive function is the difficulty of generalizing skills across different contexts. A person may effectively use a digital planner at work, where expectations are clear and reinforced, but struggle to implement the same organizational system at home, where routines are more fluid. This disconnect can be confusing and discouraging, not just for the autistic individual but also for those around them who may misinterpret their seeming nonchalance as inconsistency or lack of effort.

This difficulty arises from how different people process and categorize information differently. Without intentional guidance, strategies that are effective in one context may not automatically be effective in another context. This poses a direct challenge to self-discipline, which relies on individuals being able to flexibly apply their skills across various contexts.

Consider Andrea, a 50-year-old woman with both ASD and ADHD who excelled at structured tasks in the workplace but was impulsively overspending online, neglecting her bills and responsibilities. Although she adhered to detailed budgets at work, she struggled to replicate that discipline at home. This negatively impacted her family. However, with support, she connected her financial decisions to visual cues and established routines to pause before she spent.

In order for individuals to overcome barriers to consistently maintaining their self discipline, they often need explicit instructions, repeated practice in diverse settings, and reinforcement that emphasizes the similarities between environments. Caregivers, coaches, or therapists can facilitate this process by co-creating routines and encouraging individuals to reflect on how the tools they utilize in one domain can be adapted to suit others. Patience is crucial here—progress may be slow and nonlinear. Nevertheless, with continued effort, the ability of individuals to generalize will improve, enhancing their overall capacity to self-regulate.

Cognitive Overload and Burnout

Cognitive overload significantly undermines self-discipline by depleting the mental energy needed for decision-making, task management, and emotional regulation. Autistic adults often face substantial cognitive demands throughout the day—navigating sensory environments, masking social differences, and maintaining attention require efforts that often go unnoticed. By the end of the day, their cognitive resources may be so drained that even basic tasks like preparing meals or responding to messages become overwhelming. This creates a cycle where tasks accumulate, stress increases, and burnout follows.

In this context, burnout isn't just fatigue, it's a profound neurological exhaustion that can impair functioning for days, weeks, or even longer. When someone is in this state, the tools and habits that typically support their self-discipline may become inaccessible.

Barry, a 21-year-old college student with ASD, began to fall behind academically due to the social and sensory overload of campus life. Struggling with group work, noisy dorms, and a rigid schedule, he concealed his distress until he completely burned out and stopped attending classes. Ultimately, he dropped out of college. Isolated and ashamed, he developed depression and distanced himself from friends and family.

Barry gradually stabilized with the help of a neurodiversity-affirming therapist, a flexible part-time job, and accommodations such as online coursework and a quiet apartment. His treatment included cognitive behavioral therapy, medication for depression, and executive function coaching. A year later, he returned to school with a personalized plan: a reduced course load, asynchronous participation, and regular check-ins with a mentor. All of those strategies helped him manage his cognitive demands and rebuild his academic confidence.

Recognizing early signs of overload (irritability, procrastination, heightened sensory sensitivity, etc.) can help prevent burnout before it escalates. Preventive strategies include scheduled breaks, sensory regulation activities, reduced expectations during periods of high stress, and incorporating downtime into daily routines. Far from being indulgent, these practices are essential for maintaining executive function over time. By conserving and regularly restoring their energy, autistic adults can better uphold the habits, routines, and commitments that form the foundation of self-discipline.

Need for Predictability

Predictability and routine are more than just personal preferences for many adults with ASD—they serve as essential supports for executive function and self-discipline. Routines help reduce the unpredictability that can trigger anxiety, executive dysfunction, or emotional dysregulation. When events transpire as planned, individuals can direct their cognitive and emotional energy toward goal-oriented tasks, but when their routine is interrupted by a surprise meeting or an unexpected call, it can derail their entire day. A minor disruption can result in missed meals, forgotten items, and emotional fallout, making self-discipline extremely difficult. Establishing structured routines creates a reliable framework where self-regulation can thrive. Morning checklists, timed transitions, and rituals around meals or work can reduce uncertainty and cognitive strain. When routines are consistent, the brain spends less time figuring out what comes next, which frees up resources to complete tasks and manage emotions. Protecting routines while allowing flexibility for expected disruptions helps individuals prepare instead of react. Predictability is not only comforting, it promotes autonomy, stability, and discipline.

Consider the story of Scott, a 64-year-old man recently diagnosed with ASD after decades of struggles with work, relationships, and daily routines. For most of his life, he blamed himself for being "too sensitive," "rigid," or "unreliable," unaware that his need for control and predictability stemmed from his neurobiology. Sudden changes in schedule, spontaneous plans, and even travel triggered intense anxiety and shutdowns, which were perceived by others as avoidance or stubbornness.

After his diagnosis, though, Scott finally had a framework to understand his past. He started working with a therapist to introduce planned unpredictability into his week, such as varying his walking route or changing the time of his afternoon tea. Over time, these small shifts built his tolerance for flexibility without him having to dismantle his core routines, and for the first time, he was also able to view his structured tendencies with compassion rather than shame. By learning to create a more flexible structure, he increased his independence and deepened his relationships; his days allowed for spontaneity without fear. His journey highlights how understanding predictability as a functional need and *not* seeing it as a flaw can empower long term self-discipline and personal growth.

Why Routines and Habits Matter

Routines and habits are powerful tools for fostering self-discipline. For one thing, they provide a sense of structure and predictability, which can be especially beneficial for individuals with ASD. When our day is organized, we have fewer decisions to make, and that reduces stress and frees up our mental energy, allowing us to focus on our goals.

Habits are actions that are repeated so often that they become automatic. (For example, brushing our teeth every morning and evening is a habit we perform without thinking.)

Developing positive habits can greatly enhance our self-discipline! We can begin by focusing on one small habit at a time, such as establishing a regular bedtime, organizing our workspace, or making a daily checklist. Over time, these habits establish a framework that supports our goals and offers stability.

Routines also provide comfort, especially when we're managing unfamiliar or stressful situations. They ground us; they help us feel secure and confident even when other areas of our life seem uncertain. When routines and habits align with our goals, they become powerful allies on our journey toward personal growth.

However, relying on routine can sometimes go too far, especially for autistic individuals. The same structure that brings comfort can also become a source of anxiety when it's disrupted. At times, individuals may feel the need to control every aspect of their day to maintain stability, an urge that leads to rigid or obsessive behaviors. For example, someone might create a hyper-detailed daily schedule with no room for flexibility. If even one task takes longer than expected or is interrupted, that can result in intense frustration, derail the rest of the day, and lead to emotional distress. Instead of routines serving as tools for empowerment, they can become traps of perfectionism and anxiety.

Recognizing this potential pitfall is essential. Routines should serve us, not control us! Incorporating some flexibility and learning to accept minor deviations can help individuals maintain the benefits of having a structure without allowing the structure to become a new challenge in and of itself. Balance is key—habits and routines should support well-being, not become a source of stress.

Helpful Strategies

Time Management and Prioritizing Tasks

Effective time management is an essential part of self-discipline because it enables us to prioritize tasks, manage our responsibilities, and prevent feelings of being overwhelmed. A simple yet effective strategy when it comes to managing time involves breaking tasks into smaller, more manageable steps. For instance, instead of cleaning the entire house at once, we can concentrate on tidying up one room or area, such as a desk or shelf.

Planners, calendars, and timers help us stay organized and on track. These tools provide visual reminders of what needs to be done and when. Task management apps are likewise handy for setting deadlines and breaking larger projects into smaller, more actionable steps.

Prioritizing tasks is essential for effective time management. When deciding which tasks to tackle, individuals should start with the most important or urgent ones. Finishing these tasks first can foster a sense of accomplishment, providing motivation and confidence to address the next ones. Small wins can build momentum and make even large or intimidating projects feel more manageable.

Shelly, a middle-aged woman with ASD who balances a part-time job and caregiving duties, was finding herself constantly stressed and exhausted. Her initial approach to time management involved strict to-do lists with numerous items each day, and she became overwhelmed or emotionally distressed when she couldn't complete them all. She felt defeated even on days when she had accomplished quite a bit simply because her expectations were unrealistic.

Shelly eventually realized that her method was creating more anxiety than clarity. With the support of a coach familiar with neurodivergent needs, she shifted her strategy. Instead of meticulously tracking everything in detail, she began using a digital planner to highlight just three key priorities each day. She also learned to build buffer time between tasks and to incorporate visual reminders for breaks. This allowed her to maintain her focus without feeling overburdened, and she became more forgiving of deviations in her schedule. By adjusting her system to align with her cognitive style, Shelly transformed daily chaos into manageable routines and regained confidence in her ability to cope, ultimately reducing stress and improving her overall well-being.

Managing Strong Emotions

Emotions such as frustration, anger, or anxiety can sometimes feel overwhelming and can make it difficult to stay focused or think clearly. Self-control helps us manage these emotions and prevents them from overtaking us. When we experience strong emotions, pausing and taking deep breaths can help us regain our sense of calm. Simple techniques like counting to ten or taking a short break can provide the space we need to think more clearly.

Understanding emotional triggers is a crucial step. Triggers are situations, environments, or interactions that provoke strong reactions. For example, if loud noises make someone feel stressed or overwhelmed, they can use noise-canceling headphones to reduce the impact of those triggers. Identifying and addressing triggers allows individuals to better prepare for or even avoid overwhelming situations.

Creating a plan to manage emotions as they arise can also be beneficial. Techniques such as someone journaling their thoughts, discussing them with a trusted friend or family member, and practicing mindfulness exercises can provide relief. Individuals should experiment with different methods to discover what works best for them. Over time, these techniques can become essential parts of their emotional toolkit, helping them navigate even the most challenging situations.

Learning from Mistakes

As we noted in Chapter Eight, mistakes are a natural part of life. Not only that, but mistakes provide valuable opportunities for learning and growth. Instead of feeling discouraged, it's important to view mistakes as opportunities to reflect and adapt. Questions to ask ourselves when we've experienced a setback are "What can I do differently next time?" and "What lesson can I learn from this experience?".

When we do make a mistake and are learning from it, it's essential to be kind to ourselves. Everyone experiences moments when things don't go as planned! Viewing these moments as occasions for learning rather than as setbacks is critical. We should celebrate our efforts and progress no matter how small they may be. Growth is a process, and each step forward brings us closer to achieving our goals.

Mistakes can provide valuable insights into habits and behaviors. For instance, if an individual often misses deadlines, they may need to engage in more effective time management and planning. Insights like this can be used to develop new strategies that will promote success. Ethan is an excellent illustration of this. He's a young executive with ASD who works in a high-pressure corporate environment. Early in his career, he struggled with perfectionism and social misunderstandings—he would often replay his mistakes in his mind for days. A single miswritten email or missed meeting would lead him to self-blame and ruminate. But gradually, with the help of a mentor and a therapist, Ethan learned to reframe his thinking. He maintained a "lessons learned" journal where he transformed each mistake into a constructive takeaway. He practiced positive self-talk and allowed himself to embrace mistakes as part of his learning curve. This shift in mindset not only reduced his anxiety, it also made him more adaptable and resilient. By learning to forgive himself, Ethan became a more confident leader, one better equipped to grow from experience than be weighed down by it.

Remaining Motivated

Motivation plays a crucial role in self-discipline, especially when tasks become challenging or unpredictable. It's not just helpful, it's essential for maintaining momentum and staying committed to long-term goals. But sustaining motivation requires more than willpower! It starts with discovering personal significance in our efforts. When we link our daily actions to something deeply important—such as preserving our mental well-being, establishing a stable routine, or fostering relationships—our commitment helps us become more resilient.

For instance, enhancing time management skills involves more than merely checking items off a list. It's about alleviating stress, gaining control over our surroundings, and making room to appreciate what genuinely matters, such as pursuing a hobby, taking care of our health, or spending quality time with loved ones.

One of the most overlooked truths is that *slow is fast enough*. Sustainable progress often results from steady, consistent action rather than rushing toward a goal. Taking one meaningful step daily builds lasting habits far more effectively than burning out from overexertion does. Dedicating just ten minutes each day to decluttering a room or practicing mindful breathing can create significant positive changes over time, for example.

Rewarding ourselves for progress—no matter how small!—is another key to maintaining our motivation. A brief walk, a favorite snack, or even a quick break to engage in something enjoyable can reinforce our efforts and enhance the overall process. These moments of recognition help sustain our energy and affirm that our work is worthwhile.

Surrounding ourselves with supportive people also enhances motivation. Sharing our goals with a trusted friend, mentor, or family member can offer encouragement, fresh perspectives, and a sense of accountability. Whether it's a quick check-in message or a weekly conversation, knowing that someone is rooting for us can reignite our drive and remind us that we're not alone on the journey to personal growth.

Taking Control of Our Life

Self-discipline and self-control empower us to take charge of our lives. They allow us to make decisions that align with our values, pursue meaningful goals, and navigate challenges with confidence. By cultivating these skills, we can develop a sense of independence and self-assurance that enhances our personal and emotional well-being.

Because building self-discipline is a gradual journey, it's best to begin by focusing on one small change at a time. Whether that's establishing a daily routine, managing emotions, or improving time management skills, each step brings us closer to the life we envision. Let us not forget to celebrate our progress and trust in our ability to grow and succeed!

Five Key Takeaways

Self-Discipline is a Process—It's not About Perfection

Self-discipline develops through consistent habits and deliberate effort, not from being perfect. For autistic adults, consistency and support systems are essential. Even small routines can build confidence and reduce feelings of overwhelm. It's not about avoiding mistakes but rather learning from them and making daily progress toward stability, independence, and personal growth.

Executive Function Skills Support Self-Regulation

Executive function skills like planning and organizing are essential for self-discipline. Although difficulties in these areas can hinder follow-through, external tools (i.e., visual schedules, digital and physical reminders) can provide valuable assistance. These supports alleviate cognitive strain, making it easier to complete tasks and maintain consistency. Working *with* the brain rather than *against* it is key to developing self-control.

Emotional Control Protects Progress

Strong emotions can derail even the best intentions! That's why for individuals with ASD, regulating emotions is essential to maintain focus and consistency. Techniques such as scripting, deep breathing, or taking sensory breaks can help manage reactions and foster calmer decision making. With practice, these strategies create space for thoughtful responses and gradually enhance self-control.

Predictable Routines Reduce Stress

Routine minimizes uncertainty and liberates mental energy for making deliberate choices. For autistic adults, disruptions can be destabilizing; however, having a consistent structure provides grounding and clarity. Daily rituals like checklists and specific transitions foster stability. Predictability offers comfort and enables more focused, disciplined actions throughout the day.

Reflection Builds Resilience

Learning from mistakes instead of punishing ourselves for making them strengthens self-discipline. When things go wrong, reflecting on the situation and adjusting our strategy encourages personal growth. It's important to celebrate progress, even minor victories, and to practice patience. With kindness and curiosity, reflection becomes a powerful tool for long-term self-control and development.

Self-Guided Activities to Build Self-Discipline and Control

Activity 1: Building a Micro-Routine

Select a chaotic time of day—perhaps mornings or bedtime—and create a three-step routine. Start with one organizing action, one grounding habit, and one small self-care step. Practice this micro-routine daily for a week and observe its effects on your focus, energy, and mood.

Remember, routines build momentum through repetition.

Activity 2: Creating an Emotional Regulation Plan

Consider a situation that often causes you stress. Describe what usually happens, then create a calming script or action you can use instead, such as taking a breath, stepping away, or using a grounding phrase. Practice this response regularly so that it feels more natural to you when your emotions are heightened.

Activity 3: Generalizing One Routine

Choose a beneficial routine you already use, like writing to-do lists at work, and try to apply it in a different environment, such as at home. Adjust it slightly to fit the new context. Reflect on what feels different or more complicated and what enhances the effectiveness of the list-making process. Learning to transfer skills from one setting to another encourages flexibility and consistency.

Activity 4: Tracking Energy and Overload

Over the course of several days, track when you feel focused and when you feel drained. Look for patterns—do specific activities or times leave you feeling overwhelmed? Adjust your schedule to include rest breaks or plan to tackle more complex tasks during your peak energy periods. Effectively managing your energy will protect your ability to stick to your goals and routines.

Activity 5: Making a Motivation Board

Create a visual board (either digital or physical) that features images, quotes, or reminders that embody your goals and values. Look at it whenever you feel stuck or unmotivated. This practice will help you reconnect with your "why" and maintain a clear focus. Motivational visuals serve as effective tools for sustaining momentum and fostering self-discipline.

These takeaways and exercises offer actionable strategies to help you pursue success in a way that aligns with your unique strengths and innate talents.

11

Maintaining Resilience: Creating a Sustainable Resilient Lifestyle

Resilience isn't just something we build once and leave behind! It's a lifelong journey. For individuals with ASD, maintaining resilience involves engaging in daily practices, intentional routines, and meaningful connections. This chapter focuses on creating a sustainable lifestyle that supports resilience and enables all of us to adapt, grow, and thrive. Resilience isn't about avoiding challenges but finding ways to manage them with confidence and clarity. The foundation of resilience lies in small, consistent actions. Just as a plant grows through steady care and nurturing, resilience develops through habits and choices that support our emotional and physical well-being. While these habits may seem simple, they create a framework that empowers us to face life's ups and downs with strength and determination.

What Does It Mean to Sustain Resilience?

Sustaining resilience involves incorporating it into our daily lives. It serves as a resource during difficult times and is a fundamental aspect of our identity. It emphasizes enhancing our ability to manage stress, connect with others, and concentrate on our goals even when life feels overwhelming. Resilience is not a static trait; it's a skill we can develop and nurture over time. More importantly, resilience is about learning to cope and thrive when confronted with adversity. The skill of resilience includes the capacity to adapt, grow, and even find meaning during challenging circumstances. It means viewing setbacks as opportunities for reflection and growth, allowing us to emerge stronger and more grounded. Developing this skill requires conscious effort and self-awareness. Practices such as mindfulness, gratitude, and self-compassion enhance emotional regulation and provide a mental framework that reinforces resilience.

Additionally, resilience thrives through robust support systems—fostering and sustaining relationships with family, friends, or mentors can provide encouragement and perspective during challenging times. Participating in physical activities, prioritizing sleep, and maintaining a sense of purpose also strengthen resilience. Ultimately, the continuous process of nurturing resilience becomes a way of life—resilience is an inner strength that not only helps us endure hardships but also empowers us to flourish despite them.

For autistic adults, maintaining resilience often involves discovering methods to impose order on situations that may seem chaotic or unpredictable. Although predictability can provide a sense of security, life doesn't always follow a plan, and a resilient mindset enables people to face unexpected changes with confidence. Individuals can learn to adapt rather than feel defeated; they can take setbacks in stride and find ways to progress.

Sustaining resilience requires individuals to cultivate a deep understanding of themselves and to know that recognizing their strengths and challenges allows them to approach situations with clarity. It also involves people knowing when they need support and when they can manage situations independently. But building resilience isn't about meeting a universal standard of success, it's about each person creating a life that works for them and enables them to thrive. This process also demands patience. Growth isn't always linear—some days will feel more challenging than others. There's no need to rush or compare ourselves to others. Resilience is a deeply personal journey! The key is to concentrate on what makes us feel supported, capable, and ready to tackle challenges in our own way.

Sustaining Resilience Across Core Social-Cognitive Challenges

Sustaining resilience is a challenging journey for autistic adults; the process requires ongoing efforts across various social and cognitive domains. Individuals often face notable difficulties such as interpreting the thoughts of others, managing emotional responses, planning effectively, and navigating complex group dynamics. Areas we have highlighted throughout this book, from being able to attribute mental states to understanding nonverbal cues, represent common struggles that can impact the daily interactions and well-being of people with ASD. However, each challenge also presents an opportunity for growth! By implementing targeted strategies, seeking structured support, and engaging in self-reflection, autistic individuals can develop adaptive skills that promote emotional flexibility, social understanding, and a stronger sense of control in unpredictable environments.

The Ability to Attribute Mental States to Oneself and Others

Resilience in this area involves individuals strengthening their theory of mind—i.e., the understanding that others possess thoughts, beliefs, and emotions different from their own. For autistic adults, this can lead to misunderstandings or social fatigue, but resilience helps bridge that gap.

Consider Roy, a retired military veteran in his 60s who had never been diagnosed or treated for ASD and was working as a local handyperson. He often misinterpreted his clients' indirect comments, but with coaching and reflection over time, he learned to anticipate when someone was hinting at a need rather than stating it directly. Through daily journaling and weekly role-play sessions with a therapist, he developed a repertoire of social scripts and became more confident during social exchanges.

Gina, the college student with ASD we met earlier, struggled with group work and interpreted others' annoyance as personal rejection. However, working with a therapist helped her pause and consider alternative explanations for people's behaviors. This shift in perspective reduced her anxiety and enhanced her resilience in the classroom.

Resilience here refers to recognizing the limitations of our social intuition while developing strategies for improvement, whether through structured reflection, asking clarifying questions, or creating visual cues. As awareness increases, so does confidence in interpreting complex social environments.

The Ability to Display Emotional Reaction Appropriate to Another Person's Mental State

This form of joint emotional attention involves people aligning their emotional responses with another person's internal experience, something that often challenges autistic individuals. Saul, a man with ASD in his 40s, worked in customer service. Although he possessed technical skills, his colleagues often viewed him as cold or robotic. After attending empathy training at his workplace, however, he learned to respond with validating phrases like "That sounds difficult" whenever customers expressed frustration. Over time, these expressions became more genuine, and he was able to naturally integrate them into his interactions.

In another case, Denise, a woman in her 30s with ASD, volunteered at a community garden. She often felt unsure about how to respond when her fellow volunteers shared personal stories. By keeping a small "emotion wheel" in her notebook and journaling after interactions, she gradually learned to associate facial expressions with emotional states, which enabled her to respond more fluidly over time.

In these contexts, resilience involves continuous emotional calibrations—utilizing tools, support, and feedback to align internal emotions with external social cues even when those cues aren't instinctively understood. This kind of ongoing adjustment requires individuals to build up a heightened sense of emotional intelligence and self-awareness. It also means recognizing when emotional responses are out of sync with the environment and being willing to adapt them constructively.

Developing this level of resilience requires intentional practice, including reflective journaling, seeking constructive feedback, and engaging in open conversations with trusted individuals. Over time, this process fosters deeper emotional agility and enables people to navigate complex social situations with greater clarity and composure.

The Ability to Plan and Attend to Relevant Details in the Environment

Many autistic adults struggle with executive function, particularly in terms of attention and planning. Developing resilience in this area requires establishing external structures that support internal regulation.

Consider the life of Mary a freelance artist with ASD in her 50s. Her creativity was boundless, but deadlines often overwhelmed her. She collaborated with a coach to create a visual planning board using sticky notes and color-coded deadlines. Over time, this alleviated her panic and helped her complete projects on schedule.

Vincent, a 22-year-old warehouse worker with ASD, was learning to manage his attention by using noise-canceling headphones and a smartwatch that gave him a vibrating alert to check his task list every 30 min. These strategies helped him remain focused in a noisy and dynamic environment.

As we can see from these examples, cultivating resilience often involves implementing environmental modifications, including routines, reminders, and supportive tools to manage focus and make life more predictable.

The Ability to Understand the Communicative Content of Gaze

Understanding gaze cues—which is to say, recognizing when someone is trying to attract attention or show discomfort—is crucial for social communication and can be particularly challenging for autistic individuals.

Vern, a barista with ASD in his late 20s, frequently missed customers' visual cues, such as when they would glance toward the pastry case to indicate interest. He began rehearsing "gaze moments" using training videos and practicing them in front of a mirror, and when he was at work, instead of assuming that customers were disinterested, he would ask, "Are you looking for something specific?" Over time, this transformed his approach to customer service and helped him become warmer and more responsive in his interactions with customers.

Likewise, Marisa, an older adult with ASD who was in a book club, initially overlooked subtle glances that indicated it was someone's turn to speak. A friend privately explained these cues to her, and she began to consciously look for them during meetings. Keeping a small checklist of nonverbal cues in her planner proved to be a valuable tool for her.

As these instances illustrate, it's clear that resilience involves being persistent at learning nonverbal communication. This often happens through observation, pattern recognition, and receiving feedback from trusted allies, all of which require paying close attention to body language, tone of voice, and facial expressions while reflecting on past interactions. Over time, these insights can lead to a deeper understanding of nonverbal communications and more effective interpersonal connections.

The Ability to Work Cooperatively with Others (Joint Attention of Behavior)

Working with others toward a shared goal can be emotionally and cognitively exhausting for individuals with ASD. However, with the right structure in place, collaboration becomes possible. Lance, a 35-year-old man with ASD who was on a software development team, initially avoided teamwork. He excelled at individual tasks but found it challenging to engage in meetings. To support him, his empathic team decided to assign a "pre-brief" partner who would review tasks with him before group discussions took place. This minor adjustment alleviated his anxiety and enhanced his participation in group decisions.

Sylvia, a retired woman with undiagnosed ASD who volunteered at an animal shelter, began to face difficulties with timing during team-based care routines. Through coaching, she learned to observe group behavior using a visual flowchart of the morning cleaning schedule.

This allowed her to adjust her pace and her tasks in accordance with those of her teammates. Rather than perceiving group work as unpredictable chaos, this type of resilience is based on structure, preparation, and strategic thinking. By recognizing consistent patterns in group dynamics, individuals can foresee challenges, delegate roles effectively, and establish frameworks that encourage harmony. This proactive strategy converts disorder into manageable and collaborative momentum.

The Ability to Understand, Comprehend, Analyze, Synthesize, Evaluate and Differentiate Social Information

Processing higher-order social information, especially abstract or ambiguous cues, requires cognitive effort that can overwhelm autistic individuals. For example, when Dana, a man in his late 40s, attended workplace team building training, he found the unspoken subtexts to be overwhelming. With the help of a mentor, he started building "social flowcharts" to analyze what was said and implied and how it was received. These diagrams assisted him in distinguishing between intention and perception, making future meetings more manageable. Similarly, Sally, a 26-year-old nonprofit employee with ASD, struggled to understand social dynamics in her workplace. Through regular post-meeting debriefings with a coach, she learned to identify alliances, agendas, and tones. This reflective analysis helped her navigate what had initially felt like a maze.

In this particular context, resilience is analytical—it utilizes systems thinking, visual mapping, and metacognition to transform confusion into patterns. This approach also involves assessing the bigger picture, recognizing interconnections, and identifying root causes. By being better able to interpret complexity, individuals can develop actionable strategies and maintain clarity under pressure.

Strategies to Sustain Resilience

The Role of Daily Practices

Daily practices are essential for maintaining resilience. These small, consistent actions can profoundly impact how we feel and function each day—they ground us, helping us remain connected to our values. For autistic individuals, establishing these practices can foster a sense of stability and predictability in a world that often feels overwhelming.

Beginning the day with intention can foster a positive atmosphere. For instance, taking a few minutes to breathe deeply, stretch, or visualize our goals can help us start the day with clarity. These practices don't have to be elaborate—what matters is that they hold meaning for us. Dedicating just five minutes to doing something calming or inspiring can greatly influence how we approach the rest of our day.

Self-care is another vital aspect of daily resilience. It includes physical care, such as eating nutritious meals, getting enough sleep, and engaging in beneficial physical activity.

Emotional self-care is equally important. This may involve journaling, meditating, or simply taking a break when we're feeling overwhelmed. By prioritizing our needs, we equip ourselves with the energy and focus we need to confront life's challenges.

Reflection, too, is a powerful practice. Taking time at the end of the day to consider what went well, what felt challenging, and what we learned can help us identify patterns and areas for growth. This habit encourages us to celebrate our successes, no matter how small, and to approach challenges with curiosity rather than judgment.

While these practices are essential, it's important to remember that resilience is *not* about perfection. We do *not* need to follow our routines flawlessly to enjoy their benefits. On days when things don't go as planned, we must treat ourselves with kindness and acknowledge the effort we're putting in. Over time, these daily practices will become a natural part of our lives and provide us with a steady foundation of resilience.

Creating Meaningful Routines

Because routines provide structure and predictability—which can help alleviate feelings of anxiety or being overwhelmed—they play a vital role in maintaining resilience. For individuals with ASD, routines can be especially beneficial for fostering a sense of order and stability. A well-crafted routine serves as a roadmap that guides them through the day, making it easier to concentrate on what matters most.

Building meaningful routines begins with identifying priorities—we should all reflect on the activities and habits that help us feel grounded and productive. This may include dedicating time to self-care, pursuing personal projects, or simply enjoying a favorite hobby. Once we recognize what these positive activities are, we can consider how to incorporate them into our daily or weekly schedule.

Consistency is crucial for establishing routines, but flexibility is equally important. Life is unpredictable! It's perfectly acceptable to adjust routines as necessary. For example, if an individual's usual morning routine is disrupted, they can still find ways to incorporate aspects of it later in the day. The goal is to create routines that enhance our well-being without feeling rigid or restrictive.

Routines also help us develop positive habits over time—when we regularly repeat an activity, it becomes a natural part of our lives. For instance, setting a specific time each evening to prepare for the next day can help us feel more organized and less stressed. Over time, this habit becomes second nature, making it easier to stay on track even when life gets busy. Routines are deeply personal: what works for one person may not work for another. The key is to explore and discover what resonates with each one of us. By establishing routines that align with our needs and goals, we can foster a sense of control and purpose that boosts our resilience.

Building a Support Network

Resilience does not develop in isolation—building a support network is essential for maintaining the ability to adapt and thrive. A strong network of supportive relationships offers encouragement, understanding, and a sense of connection that can alleviate the burden of challenges.

Finding supportive relationships can sometimes be challenging for autistic individuals, but the effort is worthwhile. Support can come from various sources, including family members, friends, mentors, or professionals. It's essential to seek out individuals who respect and understand the unique needs of adults with ASD, as these relationships can offer comfort and strength.

Being part of a support network also involves contributing to it. When we offer support to others—whether by listening, sharing advice, or simply being present—we strengthen our connections and cultivate a sense of mutual trust. These acts of kindness help others and reinforce our sense of purpose and resilience.

Technology, too, serves as a valuable tool for fostering connections. Online communities, forums, and social media groups centered on ASD or shared interests offer opportunities to engage with others who relate to the experiences of autistic adults. These virtual spaces can be particularly beneficial for those who find in-person interactions challenging.

Individuals must also recognize when to seek help. Reaching out to a support network during difficult times is *not* a sign of weakness, it's a testament to resilience. Whether people need advice, encouragement, or simply someone to listen, relying on others can help them navigate challenges more effectively.

Adapting to Change

Change is an inevitable part of life, yet it can be one of the most challenging aspects of maintaining resilience. For individuals with ASD, changes in routine or unexpected events can feel especially unsettling. Therefore, developing strategies to adapt to change is a crucial skill for sustaining resilience.

One significant approach is for individuals to focus on what they *can* control, a concept we introduced in Chapter One. When faced with a new or unexpected situation, individuals should take a moment to assess their options, as breaking the problem into smaller, manageable steps can help make it feel less overwhelming. They should also identify the parts they *can* control if their daily routine is disrupted.

Preparation can likewise facilitate managing change. If individuals anticipate a change— perhaps starting a new job or relocating—it's helpful to take time to plan and gather information. Understanding what to expect can reduce feelings of uncertainty and allow individuals to approach the situation with greater confidence.

It's natural to feel uncomfortable with or resistant to change, but it's essential to approach these feelings with self-compassion. When people allow themselves to experience their emotions without judgment, that reminds them that it's okay to take time to adjust. Being patient and kind to themselves creates space for growth and adaptation.

Change can also be an opportunity for learning and growth. While it may feel challenging at times, change often offers new experiences, skills, and perspectives that enrich our lives. By remaining open to these possibilities, all of us can transform challenges into opportunities for personal development.

Sustaining Hope and Optimism

Hope and optimism serve as the foundations of resilience. They motivate us to keep progressing, even when we're faced with challenges. However, cultivating these qualities does *not* imply ignoring difficulties or pretending that everything is fine. Rather, it involves believing in our ability to overcome obstacles and create a meaningful life.

Focusing on our progress is a powerful way to cultivate hope. We should celebrate our achievements, no matter how small, and take pride in our efforts. These moments of success remind us of our ability to grow and adapt, reinforcing our belief in ourselves.

Meanwhile, optimism involves seeking opportunities even amid challenges. When faced with a setback, we can ask ourselves "What can I learn from this?" or "How can this experience contribute to my growth?" By shifting our perspective, we can uncover meaning and purpose in the toughest of moments.

Maintaining hope means surrounding ourselves with positive influences. Whether we spend time with supportive people, engage in activities that bring us joy, or seek uplifting stories, these sources of positivity can help keep us motivated and inspired.

A Lifelong Journey

Sustaining resilience is a lifelong journey that offers immense rewards. By incorporating daily practices, meaningful routines, and supportive connections into our lives, we establish a foundation for growth and well-being. As we've repeatedly emphasized, resilience isn't about avoiding challenges, it's about navigating them with strength and confidence. No matter how small, each step brings us closer to a life filled with purpose, hope, and joy.

Dean, a 52-year-old man, was diagnosed with ASD in his early 30s. For much of his life, he struggled with chronic anxiety and episodes of depression—particularly during transitions such as changing jobs or experiencing the loss of a close family member. These mental health challenges were closely linked to his sensory sensitivities and difficulties with social communication.

However, through therapy, a structured daily routine, and supportive friendships he formed in a local ASD support group, he gradually developed strategies to manage his emotional ups and downs. Dean discovered grounding in gardening, taking early morning walks, and volunteering at a community library, all activities that provided him with a sense of purpose and stability. Although he still encounters moments of overwhelm, his dedication to self-awareness and self-care continues to guide him forward. His journey serves as a reminder that resilience evolves and flourishes through compassion and intention.

Five Key Takeaways

Resilience is a Daily Commitment, not a One-Time Achievement

Resilience is not a switch to flip during difficult times—it arises from small, intentional actions practiced daily. For autistic adults, this means consistently cultivating habits that promote their well-being, whether it's a morning routine, an emotional check-in, or a moment of self-care. Sustainable resilience develops through repetition and adaptation, not perfection. Over time, these efforts can create a stable emotional foundation that helps individuals to navigate life's unpredictability with greater ease and clarity.

Emotional Awareness Must Be Cultivated

Many individuals with ASD find it challenging to read and respond to the emotions of others. Developing resilience in this area involves learning to recognize both internal and external emotional cues. People can use tools like emotion wheels or post-interaction journaling to enhance their emotional vocabulary and cultivate appropriate responses. Fostering emotional awareness creates healthier interpersonal relationships and supports a sense of belonging, both of which are critical for maintaining long-term emotional resilience and reducing a sense of feeling isolated.

Structured Routines Enable Flexibility and Growth

Although routines are often seen as rigid, they enable rather than restrict flexibility when done right. A consistent routine can ground someone amid change. For autistic adults, structured routines reduce anxiety by offering predictability while still allowing room for spontaneity. Whether it's preparing clothes the night before or using planners to visualize the day, the key is to design personalized and adaptable routines to foster resilience through order and choice.

Building Support Networks Strengthens Resilience

Resilience thrives in connection. Adults with ASD may find social interactions challenging, but having a support network—family, peers, therapists, or even online communities—makes a significant difference. Support systems provide comfort and act as feedback loops that help refine social and emotional skills. Asking for help, contributing to others, and maintaining mutual respect are all parts of a network that encourages emotional strength and provides a safe space for growth.

Cognitive Tools Can Demystify Social Complexity

Higher-order social reasoning—understanding tone, motives, or unspoken rules—can be cognitively draining. However, resilience in this area is achievable through tools like social flowcharts, visual cues, and after-action reflections. These all help to demystify situations that might otherwise feel chaotic. Autistic adults can grow their ability to navigate social environments by using these tools to gradually decode and anticipate social dynamics. This leads to greater confidence and independence over time.

Self-Guided Activities to Maintain Resilience

Activity 1: Establishing a Routine of Daily Anchors.

Choose three grounding activities—such as doing a five-minute morning stretch, reviewing your schedule, or setting a positive affirmation—and then repeat them each morning. This practice will help you establish better mental predictability and will set the tone for improved daily resilience. During the course of a week, reflect on how starting with these consistent actions has influenced your stress levels and emotional preparedness. You can adjust them as needed, but maintaining one daily anchor will strengthen your emotional resilience and organizational abilities.

Activity 2: Doing Emotion Mapping After Conversations

After a social interaction—even if it's a brief one—take a few minutes to reflect on how the other person might have felt, what cues they provided (tone, words, body language), and how you responded. Utilize an emotion wheel or draw simple smiley or frowny faces to label perceived emotions. This practice will enhance your mentalizing skills, which pertain to the ability to understand and interpret your own and others' thoughts, emotions and beliefs. Doing this for several interactions can significantly boost your empathy and lessen social misunderstandings.

Activity 3: Creating a Sensory and Attention-Planning Toolkit

Create a small toolkit with items and strategies to help you manage sensory inputs and maintain your focus. This may include using noise-canceling headphones, a fidget tool, or a smartwatch for reminders. Also be sure to organize your workspace or schedule to reduce distractions. Then set two specific goals each day that you'll help yourself achieve by using your toolkit. Every week, reflect on how this activity enhances your focus and emotional regulation skills and how your toolkit reinforces your ability to adapt to your environments (which is a key aspect of resilience).

Activity 4: Creating Visual Social Stories or Flowcharts

Select a challenging or recurring social situation—such as a team meeting or a family dinner— and create a visual flowchart or comic strip illustrating how it typically unfolds. Include key cues, emotional moments, and decision points. Next, brainstorm one or two alternative responses or outcomes. Reviewing this visual prompt before the next similar event will prepare you both cognitively and emotionally. It will also serve as a powerful resilience builder, allowing you to transform confusion into a predictable structure and giving you the confidence to engage.

Activity 5: Drilling Yourself on Your Flexibility by Changing Your Routine

Once a week, intentionally change part of your routine, like taking a different route to work or rearranging the order of a planned task. Before doing so, write about how you feel (e.g., anxious, curious, hesitant). Afterward, reflect on the experience. What went well? What felt uncomfortable? What did you learn? This practice develops resilience by gradually expanding your tolerance for change in a safe and structured manner, helping you adapt more easily when life presents unexpected challenges.

Through these insights and exercises, you'll learn effective methods for achieving success in ways that complement your personal strengths and authentic self.

12

The Path Forward: Thriving as an Adult with Autism

The journey into adulthood resembles a winding road teeming with unexpected turns, challenges, and opportunities. For autistic adults, this path may have distinctive twists that set it apart from the experiences of others. Nonetheless, it's also a road rich in potential, growth, and self-discovery. Thriving as an adult with ASD is not merely about following someone else's map, it's about individuals creating their own paths. It involves embracing their authentic self, building on their strengths, and continually learning the skills and strategies that will empower them to navigate life's complexities with confidence.

This chapter reflects on the ongoing journey of resilience, emphasizing that it's not a destination to be arrived at but rather a daily process that individuals with ASD actively engage with. Resilience is *not* the absence of difficulties or setbacks—it's about the ability to face them with determination, adaptability, and hope. It's about individuals finding meaning and fulfillment in their own progress, celebrating small victories, and cherishing the relationships and connections that enrich their lives. Thriving is a continuous, evolving process! In this chapter, we aim to inspire and guide autistic adults as they continue along their life's journeys.

Embracing Authenticity

The heart of being able to thrive is for all of us to fully embrace who we are. Autism is more than just a diagnosis! It's an integral part of an individual's identity—it shapes the way they experience, understand, and interact with the world. This unique perspective is not something to hide or downplay; it's something to honor and celebrate. Embracing authenticity means recognizing that differences are not deficits but are in fact strengths that allow people to bring diversity, creativity, and value to the world.

For individuals to truly embrace authenticity, they must first understand their strengths and challenges. Acknowledging how their brain works differently can help them appreciate their individuality and discover ways to leverage their strengths. For example, an autistic adult might excel in tasks that require focus, attention to detail, or systematic thinking. These abilities could make that person a valuable contributor to technology, research, or the arts. Similarly, having heightened sensory awareness or a unique way of processing information might enable an individual to notice patterns or details that others overlook, in the process offering insights that enrich their personal and professional lives.

Embracing authenticity also involves rejecting societal pressures to conform to narrow definitions of success or normalcy. The world often promotes ideals that may not align with the experiences or values of individuals with ASD, but everyone must recognize that each person's path is their own and that success is personal. Thriving doesn't mean meeting someone else's expectations, it means that every individual lives in a way that feels meaningful, fulfilling, and true to them. Whether that involves pursuing a career that sparks their passion, nurturing relationships that bring them joy, or finding contentment in a quiet, structured routine, one person's version of thriving is just as valid and vital as anyone else's.

Finally, embracing authenticity requires self-compassion. All of us may experience moments of doubt, frustration, or comparison, but these feelings are part of being human. It's crucial that we treat ourselves with kindness and understanding, recognizing that growth takes time and that each person—including ourselves!—deserves patience and grace. By accepting and celebrating who we are, we can create lives that reflect our true selves and enable us to thrive.

Staying True to One's Values

Values represent the principles and priorities that matter most to us, providing a sense of purpose and direction. For autistic adults, staying connected to their values can offer clarity and stability, especially during uncertain times or changes. These values may include kindness, honesty, creativity, independence, or any other ideals.

To stay true to our values, it's helpful to regularly reflect on what those values are and how they influence our lives. We need to ask ourselves questions like "What gives my life meaning?" and "What do I want to stand for?" This introspection can help us identify what truly matters and ensure that our actions align with our priorities. For instance, if we value helping others, we might volunteer for causes that inspire us or support a friend in need. If independence is a core value, we can focus on building skills that enhance self-sufficiency, such as managing finances or learning new life skills.

But of course, living according to our values doesn't mean that life will always be smooth or free of obstacles. It *does* mean using our values as a guide to help us navigate challenges and make decisions that feel right. For example, if we encounter a difficult situation at work or in a relationship, reflecting on our values can help us choose a course that aligns with our principles. Connecting to what matters most can provide us with stability and purpose even in times of uncertainty.

Moreover, staying true to our values can help create an authentic and fulfilling life. When our actions align with our core principles, we're more likely to feel content and at peace with ourselves. This alignment fosters resilience and enables us to weather challenges with greater confidence and clarity. Regularly revisiting and reaffirming our values allows us to cultivate a life that reflects who we are and what we stand for.

Creating Supportive Environments

Thriving with autism doesn't happen in a vacuum—no matter how much internal growth individuals pursue, the environments in which they live, work, and socialize matter. A supportive environment isn't about perfection or control, it's about creating spaces that help people feel grounded, focused, and safe to be themselves. These environments can buffer stress, reduce sensory overload, and enhance people's ability to function and flourish.

For many adults with ASD, creating a supportive environment begins at home. Whether someone lives alone, with family, or with roommates, organizing their living space around their sensory needs and daily routines can make a significant difference. For example, how do lighting, noise levels, and clutter impact their mood and energy? They might find comfort in soft lighting, noise-canceling headphones, or a structured layout that minimizes visual chaos. It's important for individuals to envision their home as a personalized retreat—a place that restores their energy instead of draining it.

The workplace is another key environment where support can significantly impact the experience of autistic adults. Clear communication, predictable routines, and respect for everyone's preferred working styles can help them thrive professionally. Advocating for accommodations like flexible scheduling, written instructions, and/or a quiet workspace isn't asking for special treatment, it's creating the conditions that individuals need to do their best work. If adults with ASD are job hunting, it's wise of them to seek out organizations with inclusive cultures or disability-friendly hiring practices. These workplaces are more likely to recognize each individual's strengths and invest in their success.

Social environments are also very important. Autistic adults should seek out communities and friendships that allow them to be authentic without the pressure to mask or perform. These include online groups focused on shared interests, local meetups for neurodivergent adults, or relationships with individuals who appreciate their communication style. It's okay for individuals to step back from relationships that constantly drain them or require them to suppress their true selves.

Digital environments also matter. Consider how the digital world influences everyone's well-being, from particular ways that phones are set up to online interactions. It's advantageous for individuals with ASD to utilize apps and tools that assist them with staying organized, tracking emotions, and/or managing sensory overload. Digital spaces that cause comparisons or stress should be limited.

Ultimately, whether or not all of us have supportive environments depends on intentional choices. We may not be able to control every aspect of the world around us, but we *can* shape many elements to better suit our needs. Thriving means creating a life that supports—not suppresses—the unique way of being that each one of us has.

Redefining Social Connections

Autistic adults often experience and express social connections differently from their neurotypical peers. This isn't a flaw, it's simply a different style that deserves respect. Redefining what social connection is or should be means letting go of the pressure to conform to conventional norms and instead building relationships in genuine and sustainable ways.

For some, deep one-on-one conversations are more meaningful than group settings. Others prefer connecting over shared interests rather than making small talk. There's no single "right way" to socialize. The key is to honor what works for each person. Individuals don't have to be outgoing or highly verbal to build relationships—communication through shared activities, creative expression, or even silent companionship can be just as valid and fulfilling.

Many adults with ASD prefer fewer, more meaningful relationships instead of large social circles. This is perfectly acceptable. A strong connection with one or two people can be more enriching than being surrounded by numerous acquaintances. In other words, it's a focus on quality, not quantity.

Creating supportive social circles may involve seeking out individuals who understand neurodivergence, whether they are autistic themselves or simply empathic and open-minded.

Neurodivergent communities provide a shared experience that's difficult to find elsewhere. Autistic individuals should look for online forums, local support groups, or community centers that host autism-friendly events.

Setting boundaries is also a crucial aspect of thriving socially. We all have the right to manage our energy, choose when and how to interact with others, and say no without guilt. Not every social opportunity needs to be pursued. Learning to advocate for specific social needs— such as taking breaks, using scripts, or leaving early—helps individuals protect their well-being.

Navigating misunderstandings can be challenging, but clear communication helps. If adults with ASD feel safe doing so, they might explain their communication style to those close to them—they might say "I take longer to respond because I like to think carefully" or "I prefer texts over phone calls." The more the needs of autistic individuals are understood, the more comfortable their relationships will become.

Redefining social connection involves people reclaiming their way of relating to others. It's not about being more social! It's about achieving social fulfillment in ways that resonate with each one of us. Whether through shared routines, mutual interests, or deep conversations, the connections we make should energize—not exhaust—us.

Managing Transitions and Change

Change is hard for most people, but for adults with ASD, transitions can be especially overwhelming. Whether it's a new job, a move, a change in routine, or a relationship shift, these transitions can disrupt the structure and predictability many rely on. However, with the right strategies, individuals can manage these changes to reduce anxiety and support long-term growth.

The first step is acknowledging that transitions are challenging and that it's okay to feel overwhelmed. Feeling overwhelmed doesn't mean that individuals are failing—it just means that their brains need time to adjust, and it's crucial that they give themselves that space. Whenever possible, it helps to anticipate a change, because if people know a transition is coming, they can create a plan and then break it down into manageable steps. They can rely on checklists, timelines, or visual schedules to organize what needs to happen and when.

But preparation goes beyond logistics—it also involves emotional readiness. Autistic individuals can strive to identify their nervousness and what kinds of support will help them. Will they need extra downtime? Would discussing a given situation with someone be helpful? Anticipating their emotional responses enables them to plan how to manage those responses through coping strategies, sensory tools, and supportive conversations.

Flexibility is a skill people can build, not something they're born with or without. We can all start building flexibility by intentionally making small changes—for example, we can switch our routine slightly or engage in a new activity. Practicing adaptability in low-stress situations can help us prepare for bigger shifts. And when significant transitions happen, we can remind ourselves that uncertainty is temporary. The structure will return—we just need to give it time to settle in.

Support networks are invaluable during times of change. Adults with ASD should let others know what they're experiencing and ask for help when they need it. Even just having someone listen can make a difference. If change triggers past trauma or intense anxiety, they can consider working with a therapist who understands autism. These professionals can help them process transitions and develop strategies tailored to their needs.

Transitions are part of life, but they don't have to derail individuals' progress. By approaching these transitions with intention, planning, and support, autistic adults can navigate change in a way that respects their needs and strengthens their resilience. Growth doesn't come from avoiding change; it comes from individuals learning how to handle change on their own terms.

Practicing Self-Advocacy

Self-advocacy is a vital life skill, particularly for adults with ASD. It involves them recognizing their needs, understanding their rights, and communicating clearly to ensure they're respected. Whether they request accommodations, set boundaries, or explain their communication style, self-advocacy empowers them to take control of their lives.

One of the first steps in self-advocacy is self-awareness, i.e., individuals understanding what helps them function at their best and what causes them distress. Someone may work best with clear, written instructions, or they may require extra time to process information. Loud environments may wear them down; spontaneous changes might throw them off. When individuals identify their preferences, they can confidently ask for what they need. Next is communication. Autistic adults don't have to be forceful or extroverted to advocate for themselves, but they do need to be clear and direct. Someone might say "I process information better when I can write things down" or "I'd like advance notice before the schedule changes." Practicing scripts for everyday situations can be helpful, especially when individuals are anxious or under pressure.

Self-advocacy also involves understanding the legal rights of individuals. In many countries, laws exist to protect people with disabilities in workplaces, schools, and public services. Individuals with ASD should learn about the accommodations they're entitled to and how to request them. Resources such as advocacy organizations or disability rights centers can provide guidance and support.

Boundaries are another crucial part of self-advocacy. It's okay to say no, ask for space, or tell people when something isn't working—no one owes anyone else an explanation for protecting their peace. When people say, "I need time alone right now" or "That tone feels dismissive," that represents them standing up for their well-being.

A core aspect of self-advocacy to remember is that it's a skill people develop over time. No one always gets it right, and that's fine. Each experience teaches individuals something new that will help them grow more confident in expressing their needs and navigating challenges with.

practice. Moreover, the more autistic adults advocate for themselves, the more they encourage others to do the same. When they speak up—not just for their legal rights, but for their right to thrive—they're not merely surviving. They're shaping a life where everyone can truly belong.

Developing One's Strengths

Every autistic individual possesses unique strengths and talents that can be nurtured and developed. These abilities are tools for success and are sources of confidence, creativity, and fulfillment. Identifying and building on personal strengths is a powerful way to thrive personally and professionally.

Discovering personal strengths often involves exploration and self-reflection. Individuals should consider the activities or tasks that come naturally to them or that bring them a sense of accomplishment. Someone may excel at solving complex problems, for example, or have a gift for artistic expression, or possess a knack for organizing and planning. These strengths can guide individuals toward opportunities that align with their abilities and interests—a love of numbers might lead to a career in finance or data analysis, while a passion for storytelling might inspire someone to write or create content.

Once individuals have identified their strengths, they must prioritize practicing and developing those strengths. Engaging in activities that are challenging and that further refine their abilities can help build confidence and open new doors. Adults with ASD shouldn't be afraid to try new things or revisit hobbies they enjoyed in the past! Each experience adds to their skills and helps them discover what they want. Growth is gradual; even small steps can lead to significant progress over time.

It's also important to recognize that individuals' strengths can evolve. Challenges they face today might become areas of expertise tomorrow as they develop strategies and gain experience, and similarly, the strengths they build now can serve as a foundation for future opportunities. Thriving is a dynamic process. When individuals invest in their abilities, they can create a life that reflects their potential and aspirations.

Building Resilience as a Lifelong Process

Resilience is not an innate trait but a skill that can be cultivated and strengthened over time. For autistic adults, resilience involves finding ways to adapt to life's challenges, manage stress, and maintain emotional and mental balance. It's about recognizing that setbacks are a natural part of life and viewing them as opportunities for growth rather than as insurmountable obstacles.

Self-care, too, is a cornerstone of resilience. For all of us, taking care of our physical and emotional well-being provides the energy and focus we need to face life's challenges. This includes maintaining a consistent sleep routine, engaging in regular physical activity, and practicing mindfulness or relaxation techniques. These habits create a foundation of stability, helping us approach difficulties with a clear mind and steady resolve.

Another critical aspect of resilience is building and maintaining a support network. As much as possible, we should surround ourselves with people who understand, respect, and encourage us. Whether these connections are with family members, friends, or professionals, they provide a sense of belonging and a source of strength during difficult times. And we shouldn't hesitate to seek help when we need it—doing so is essential to becoming more resilient. It is *not* a sign of weakness.

Resilience also involves cultivating a growth mindset. Instead of fearing mistakes or failures, adults with ASD can begin to embrace them as opportunities to learn and improve. This requires that individuals reflect on what went wrong, identify what they could do differently, and apply those lessons. This approach builds confidence and enhances their ability to adapt to and thrive in new situations.

Journeys with ASD

Thriving as an autistic adult is a journey of continuous growth, self-discovery, and resilience. It involves individuals embracing their authentic selves, remaining true to their values, and developing the skills and strengths they need to navigate life's complexities. Although the path may not always be easy, it *is* filled with opportunities for fulfillment, joy, and achievement. As adults with ASD move forward, it's important for them to remember that thriving isn't about achieving perfection but about making progress and celebrating each step. Every challenge overcome, every goal reached, every connection nurtured is a testament to their resilience and determination. A key part of their journey is to trust in their ability to navigate the road ahead and to know that their journey is uniquely *theirs*.

Living with ASD may come with obstacles—social expectations, sensory overload, pressure to mask—but none of these define people with ASD. What defines them is their ability to keep going, to learn and adapt, and to find ways of existing that feel right to them. Thriving means setting boundaries that protect their well-being, seeking out communities where they feel understood, and allowing themselves to say no to what drains them.

Autism isn't a limitation, it's a different way of experiencing the world. With the proper support, tools, and mindset, individuals with autism can build a fulfilling life that works for them *and* allows them to thrive because of who they are. Whether it's through a career tailored to their strengths, friendships grounded in mutual respect, or creative pursuits that bring them joy, there's no singular path to success. Success is personal and may look different for each person—and that's perfectly okay.

Autistic individuals can create lives that reflect their authenticity, values, and aspirations. They can embrace the opportunities that come their way and let their strengths guide them toward a future filled with possibilities. They have the right to dream big, pursue their version of happiness, and take pride in how far they've come. The path forward is theirs to shape, and the options are endless.

Five Key Takeaways

Embracing One's Authentic Self is Empowering

Thriving with autism begins with individuals acknowledging that their brain works differently— and that's not a flaw. It reflects their unique way of thinking, feeling, and experiencing the world, adding value and insights. They don't have to fit into neurotypical molds to succeed. By honoring their own needs, communication styles, and sensory preferences, they lay the foundation for true confidence and growth. Being themselves isn't a compromise—it's a strength that fosters lifelong fulfillment.

Living by One's Values Creates Meaning

Values are like a personal compass—they guide our decisions, protect our energy, and help us focus on what truly matters. When life feels uncertain, reconnecting with our values brings clarity and purpose. They enable us to choose relationships, jobs, and environments that promote our growth. Whether individuals with ASD prioritize honesty, creativity, or independence, staying grounded in their core beliefs ensures that their actions align with their authentic self and support their long-term well-being.

Supportive Environments Make a Difference

No one thrives in isolation. Physical, emotional, and digital environments are critical for feeling safe and successful. Thoughtful adjustments—such as reducing sensory overload at home or advocating for accommodations at work—help alleviate stress and enhance daily comfort for autistic adults. Similarly, surrounding themselves with people who respect their communication styles and support their goals reinforces their sense of belonging. Individuals with ASD don't have to change who they are—instead, they can change the environments that surround them.

Social Connections Can Be Redefined

Autistic individuals don't need to socialize in conventional ways to have meaningful relationships. Whether they bond over shared interests, engage in nonverbal companionship, or cultivate one or two deep connections, their version of social fulfillment is entirely valid. Regarding friendships, they should reject the idea that "more is better" and instead focus on quality, honesty, and comfort. Building social connections that fit their needs will make it much easier to engage, recharge, and feel truly seen.

Resilience is Built, not Born

Resilience is not about avoiding hard times but about learning how to adapt when they come. This includes practicing self-care, seeking support, and giving ourselves permission to slow down. Over time, small efforts to manage stress, reflect on setbacks, and continue moving forward build strength. Establishing consistent routines, maintaining boundaries, and leveraging strengths are all acts of resilience. The more all of us practice doing these things, the more confident and prepared we'll become.

Self-Guided Activities to Thrive as an Adult with Autism

Activity 1: Designing Your Personal Values Map

Create a simple visual map of your values. Start by writing five to seven words that represent what matters most to you (e.g., honesty, freedom, kindness). For each word, craft a sentence about how it manifests in your daily life. Next, reflect on ways that you can align more closely with each value. This exercise will clarify your priorities and provide a tangible tool to reference when facing decisions or transitions.

Activity 2: Doing a Sensory-Friendly Space Audit

Take a tour of your home or workspace and note how each area makes you feel—calm, anxious, overstimulated, etc. Consider the lighting, noise, organization, and comfort levels. Identify three small changes you can make to reduce sensory stress, such as adding soft lighting, using noise canceling headphones, or reorganizing clutter. This audit helps you intentionally shape your environment to better support your energy, focus, and well-being.

Activity 3: Building a Personal Self-Advocacy Script Library

Create five to ten scripts for common scenarios where you may need to advocate for yourself, such as "I need written instructions" or "Can I have a few extra minutes to respond?" Practice articulating them aloud or in writing. You can even role-play the scenarios with someone you trust. Having these scripts prepared helps reduce anxiety in the moment and builds confidence in your ability to communicate clearly and assertively.

These insights and exercises will show you practical ways to reach your goals by working with your natural abilities and staying true to who you are.

13

Epilogue

The following story represents a collection of experiences we've heard from many autistic individuals whom we've worked with over the years. Each struggle, insight, and moment of resilience reflects real stories, challenges, and triumphs. This narrative closely mirrors the life of one particular individual—a man who, like many others, navigated life without the knowledge or support that a diagnosis might have offered. His journey is one of perseverance, adaptation, and eventual self-acceptance, illustrating both the personal and universal aspects of living with undiagnosed ASD.

Looking Back: A Life Lived on the Spectrum

I turned 60 today. It snuck up on me, though in some ways, I've been waiting for it my whole life, not because of the number, but because of what it brings—a moment of reflection, a moment of understanding.

For most of my life, I moved through the world like a visitor in a foreign country. The rules of social interaction, the expectations, and the unspoken cues were like an intricate language I never quite mastered. It wasn't until a few years ago, deep into adulthood, that I found out why: I have autism spectrum disorder.

No one told me. No doctor diagnosed me as a child. No teacher suggested it. No well-meaning therapist gave me the tools to understand my mind. I grew up being told I was odd, quiet, blunt, and obsessed with small things. I was either too much or not enough—never just right.

Childhood: The Silent Struggle

I remember being five years old, sitting in my kindergarten classroom, staring at how the dust swirled in the sunlight. The other children were playing, laughing, and making up imaginary games. I was happy in my world of patterns and details, but that wasn't what they wanted from me. The teacher nudged me toward the others, urging me to "go play," as if it were as simple as deciding to walk through an open door. But there was no door for me—just a wall between their world and mine.

The kids didn't dislike me, exactly; I wasn't bullied, at least not much. But I wasn't understood, either. I'd say the wrong thing or nothing at all. I couldn't tell when a game was over or a joke was made at my expense. So I stayed in my own space, keeping close to my books and routines. The real world was unpredictable, but *my* world had structure.

My parents worried. They noticed how I struggled to connect and did their best to help. "Just talk to them," they would say, as if it were that simple. They called me "difficult" when I had a meltdown over a change in plans. When I lined up my toys in a specific order and panicked when they were moved, they labeled me as being "stubborn." They didn't know any better, and neither did I. I just learned to keep it all inside.

By age seven, I had begun to notice my differences more clearly. I loved memorizing facts and spent hours reading books about space, dinosaurs, and the ocean. Other kids talked about cartoons and superheroes—things I didn't understand or care about. I preferred the swing set at recess because it had a rhythm I could control. The motion soothed me when the classroom felt too loud or chaotic. But when the other kids ran off together, forming groups with unspoken rules, I stayed behind, never quite knowing how to invite myself in.

One time, a group of classmates played a game of make-believe in which they were explorers. I wanted to join them, but I didn't know how. I just stood nearby, watching, absorbing their movements and words, waiting for a cue that never came. When one of them finally noticed me, they asked, "What are you doing?" I didn't have an answer. I shrugged and walked away, retreating into my world again.

By age ten, I had developed strategies to cope with the confusion of social life. I mimicked what I saw on TV and rehearsed phrases before speaking. Sometimes that worked, but other times it felt forced and unnatural. School had become more structured, which was a relief, but group projects remained challenging. I wanted to do things my way, to follow the rules as I understood them, and it frustrated me when others deviated from the plan.

One afternoon, our class was given an art assignment to draw our "perfect world." While others sketched fantastical lands with castles and talking animals, I carefully designed a city with neat streets, perfectly aligned houses, and a predictable weather pattern. When a classmate looked over my shoulder, they laughed and said, "That's so boring! Where's the fun?" I didn't understand what was wrong with it. To me, the order was beautiful.

Looking back, I know that I wasn't broken or difficult. I just saw the world differently. And while I didn't always fit in, my way of thinking gave me a different kind of strength—one that I would come to understand and appreciate more with time.

Teen Years: The Masking Begins

When I entered middle school, I knew the rules well enough to get by. I observed people like a scientist, figuring out when to nod and when to laugh. I mimicked how others spoke and moved, hoping no one would realize I was navigating every interaction by guesswork. It was exhausting. Every school day drained me, but I persevered. My parents didn't understand why I was so tired all the time. "You don't play sports, you don't hang out with friends—how can you be tired?" they would ask me. But back then, I lacked the words to explain what masking did to a person.

Social gatherings were the worst—loud, crowded, full of conversations that moved too fast for me to follow. I started avoiding them, preferring the comfort of solitude. My parents worried, of course. "You need to make an effort," they said. "You can't go through life alone." So I tried harder. I went to events, stood at the edges of conversations, and pretended to enjoy them. I smiled, nodded, and laughed in the right places. And I went home feeling like I had run a marathon.

Dating was another minefield. I watched how others flirted and mimicked how they teased and joked, but I never quite understood the unwritten rules. I had a couple of short-lived crushes, but the idea of navigating a relationship—constantly having to perform—felt overwhelming. I went to a few school dances; each one was a blur of flashing lights, pulsing music, and strained small talk. I rarely danced, except when someone pulled me onto the floor, and even then, I moved stiffly, hyper-aware of every awkward step.

Then there was prom. My only prom. I went because I thought I was supposed to, and besides, it seemed like something I'd regret skipping. I rented a tuxedo that felt stiff and unnatural; the bowtie was an irritating pressure around my neck. I asked a girl to be my date—a quiet, bookish girl, someone who, in hindsight, was probably masking just as much as I was. She said yes, maybe because she felt the same unspoken pressure to participate.

We matched in an accidental way—she wore a simple dress, nothing flashy, and didn't seem interested in the elaborate traditions that others obsessed over. We took the required photos, stood awkwardly with a group of classmates, and exchanged forced smiles. The dinner was a minefield of small talk, with both of us struggling to keep up with the quick, overlapping conversations at the table. When we danced, it felt hesitant, almost mechanical. We swayed more than we moved, both of us uncomfortable with the unspoken rules of proximity and rhythm. She wasn't much for crowds either, so we spent most of the night at the edge of the room, watching rather than participating. I think she was relieved when I suggested leaving early. We grabbed fast food on the way home, then sat in the car and ate in comfortable silence. In the end, prom wasn't magical or life-changing, but it was tolerable, maybe even lovely in its own way. It was one of the few times when I didn't feel utterly alone in my awkwardness.

College Years: The Balancing Act

College was a fresh start, or at least that's what I told myself. No one knew me there. I could reinvent myself and become the person I had spent years pretending to be. But masking in high school had been exhausting, and college only amplified the pressure. The social rules were looser but still present, and they shifted depending on the situation—classrooms, dorms, parties, study groups. I spent my first year figuring out where I fit, which usually meant keeping my head down, speaking only when necessary, and mirroring the people around me.

Academically, I thrived. College provided structure, clear expectations, and a level of independence I had never experienced before. I could schedule my days in a way that made sense; I could spend long stretches alone in the library, deeply focused on my coursework. However, the social expectations didn't vanish. Group projects, casual hallway conversations, the dreaded small talk before class—I still had to perform. I went to a few parties, stood at the outskirts as always, and tried to drink just enough to blend in but not so much that I would lose control.

By the time graduation was nearing, I had learned to balance things better. I still needed solitude to recharge, but I had also figured out how to navigate social settings with less strain. The biggest challenge came next: job hunting.

The job interviews were brutal. Every one of them felt like it required an advanced level of masking—it was a performance where I had to convince strangers that I was confident, personable, and precisely what they were looking for. The unwritten rules of socializing had evolved into something even more intricate: networking, office politics, and professional small talk. I rehearsed answers, studied body language, and did my best to sound natural. Eventually, I landed a job, but the relief was short-lived. I had succeeded in getting through the door, yes…and now I had to survive the daily interactions, meetings, and workplace dynamics. College had been challenging, but the real world had its own rules—ones I was still trying to decode.

Adulthood: A World Built for Others

By my 20s, I had mastered the art of getting by. The job I landed was in data analysis. Numbers made sense to me—they had rules and patterns. People, on the other hand, remained unpredictable.

Work wasn't easy. Office politics, unwritten expectations—I got in trouble more than once for saying something "the wrong way." If a coworker asked for my opinion, I gave it, not realizing they often wanted reassurance, not honesty. I struggled in meetings because I kept missing the subtle shifts in tone that told everyone else when to move on and when to speak up. I was competent—sometimes even brilliant—in my work, but I was never comfortable.

Relationships were harder. Dating felt like an elaborate game with no clear rules. I never understood flirting and never picked up on the signals that others seemed to recognize instinctively. More than once, I was accused of being cold and detached. But I wasn't—I felt deeply. I just didn't know how to express it in ways that others understood.

A Love That Tried, A Love That Ended

I did marry, though. Her name was Anna. She was patient and kind, a woman who saw something in me that others didn't. While some found my intensity overwhelming, she admired my focus, honesty, and dedication to what I cared about. But even she struggled with the way my mind worked.

"Sometimes it's like you're in another world," she told me once. She wasn't wrong. There were days when I felt disconnected, my thoughts lost in ideas and projects that consumed me. I loved Anna, but love alone wasn't always enough. We lasted ten years before she left. "I need more from you," she said. And I didn't know how to give it.

But from our time together came the greatest part of my life: our daughter, Lily. From the moment she arrived, my world shifted in ways I hadn't expected. She was small and delicate, yet so full of life and curiosity. She inherited her mother's patience but carried my intensity—a quiet determination that reminded me of myself in both comforting and unsettling ways. Parenting didn't come naturally to me. I struggled with the unpredictability and emotional depth required to nurture a child as effortlessly as Anna did. But I tried. I read to her every night, even when my mind was elsewhere. I took her to the park, even though social settings drained me. I loved her fiercely, even when I didn't always know how to show it. When Anna and I divorced, I thought Lily would leave me, too—but she didn't. I braced myself for the inevitable distance, the slow unraveling of our bond, the way children sometimes drift toward the parent who provides more stability, more warmth. But Lily stayed. Not in the literal sense, of course—she lived with Anna during the week—but in the ways that mattered most. Our visitations on weekends became a sacred time, something I clung to even when I wasn't sure I was getting this whole parenting thing right.

At first, I worried that my structured, inward-focused world would bore her and that she'd rather be anywhere else. But she surprised me. She'd show up on Friday evenings with her overnight bag slung over her shoulder, always wearing that quiet smile, as if she were stepping into a space where she felt safe and understood. We spent Saturdays walking in the park, with me pointing out birds and explaining how trees change with the seasons while she listened intently. At night, I'd read to her—sometimes the same stories over and over because she loved their familiarity.

She never asked for more than I could give, never demanded that I change into something I wasn't. And yet, over time, I did change in small ways. I learned to be more present, listen more carefully, and meet her gaze instead of letting my thoughts drift elsewhere. I realized that Lily didn't need me to be like Anna—warm, expressive, overflowing with affection. She just needed me to be there. And so, I was.

Middle Age: The Unraveling

By my 40s, the effort to mask had taken its toll. Anxiety and depression—I had both, though I didn't recognize their names until much later. Work became unbearable, and I entirely avoided social situations. My world shrank to routines and rituals. They provided comfort, but they also trapped me.

It was around this time that I started to suspect there was something different about me. The internet had grown into a space where people shared experiences, and one day, I stumbled across a forum for autistic adults. I read their words and saw myself. The sensory overload, the social confusion, the need for structure…they were all there.

I took an online test, and then another. Each one produced the same result—the numbers, categories, and explanations all pointed in one direction. The tests appeared to confirm that I had ASD. I wasn't sure how to feel. Were these just internet quizzes, or were they revealing something I had overlooked for years? I started researching, comparing symptoms, and reflecting on past experiences. The way I struggled with social cues, routines, and sensory overload suddenly made more sense. The more I read, the more it felt like a missing piece of my identity had finally clicked into place.

At first, I felt a strange mix of relief, confusion, and even disbelief. Could this be the answer to why I had always felt different? Why social interactions had often felt like trying to decode a foreign language? Why certain noises, textures, or changes in routine had unsettled me in ways I couldn't explain? I had spent my whole life not knowing who I was, shaping myself around the expectations of others, masking my struggles so well that even *I* hadn't recognized them for what they were.

Looking back, moments from my past suddenly made sense—why I had always been drawn to routines, why specific conversations had drained me, why I had replayed social interactions in my head, endlessly analyzing every detail. The frustration I had felt for years wasn't a personal failing, it was part of how my brain processed the world.

Yet, even with this newfound understanding, questions flooded my mind. What did this mean for my future? How would this change the way I saw myself? Did I even want to be labeled? But one thing was sure—these weren't just test results. Their collective answer was a key to unlocking a part of myself I had never fully understood.

Acceptance: The Road Home

I didn't receive an official diagnosis until I was 53. Having my entire life summarized in a few clinical terms felt strange: "autism spectrum disorder." It wasn't a label I had ever pursued, yet when I heard it, something clicked into place. The confusion, exhaustion, and lifelong feeling of being out of sync with the world suddenly made sense.

Getting to that point, however, was not easy. I had hesitated to seek an evaluation, fearing the psychologist might dismiss me as merely eccentric or socially awkward. I had spent decades adapting, masking my differences, and trying to fit in. What if I had done it too well? What if I was just a bit odd and nothing more? That fear almost kept me from making the appointment. But I was fortunate. The psychologist I saw wasn't just any clinician, he had written three textbooks on autism in children and adults. He understood the subtleties that many other professionals had missed. The evaluation process was thorough—structured interviews, developmental history, and cognitive assessments—all aimed at piecing together a lifelong pattern of behavior. He didn't just listen to what I said; he *saw* the things I had spent a lifetime hiding.

The anger came first—anger over the years of struggle and lost opportunities for support. But then relief followed. It wasn't my fault that I didn't conform. The world had never been designed for people like me.

And then a gradual transformation began. I started to understand myself in ways I never had before. I gave myself permission to stop trying so hard to be someone I wasn't. I allowed myself to rest and recharge without guilt. I connected with others in the autism community, who understood me like no one ever had. For the first time, I felt a sense of belonging.

Sixty: A Life Reclaimed

And now, I'm here: I'm 60. Looking back, I see a life full of struggles, yes, but also resilience. I survived in a world that never made room for me. I adapted, even when it exhausted me. I learned, even when no one taught me.

I don't regret my life. But I wish I had known sooner that I wasn't broken, just different. If I could talk to my younger self, I'd tell him: "You're not alone. You never were. You don't have to conform to the world's expectations. You just have to be you."

And that? That's enough. More than enough, in fact. It's everything. Learning to embrace my differences instead of fighting them has been the most liberating part of this journey. I no longer feel the need to apologize for who I am, nor do I seek validation in spaces that refuse to understand me. I have discovered my own kind of peace—one built on self-acceptance, patience, and the realization that my way of being in the world is just as valid as anyone else's.

Nowadays, I choose authenticity over approval. I prioritize my needs, my comfort, and my happiness without shame. I have reconnected with my daughter and have made up for lost time by expressing love in ways that feel natural to both of us. I have built friendships that demand no performance but offer genuine connection. I have carved out a life that finally feels like mine.

As I step forward into the years ahead, I do so with gratitude—not just for my resilience but also for the researchers, advocates, and professionals who refused to accept the status quo. They dedicated their lives to studying autism and fought for a world where individuals like me could finally find understanding. Their efforts have enabled others to receive the support I never had, and I am profoundly thankful for that.

Resources

Books for Adults with Autism

1. *The Journal of Best Practices* by David Finch (personal memoir on marriage and autism)
2. *Unmasking Autism* by Devon Price (explores masking and self-acceptance)
3. *NeuroTribes: The Legacy of Autism and the Future of Neurodiversity* by Steve Silberman
4. *Life on the Autism Spectrum* by Dr. Matthew Bennett (practical advice)
5. *Living Independently on the Autism Spectrum* by Lynne Soraya
6. *Look Me in the Eye: My Life with Asperger's* by John Elder Robison (personal perspective)
7. *The Asperger's Syndrome Workplace Survival Guide* by Barbara Bissonnette
8. *The Hidden Curriculum for Understanding Unstated Rules in Social Situations* by Brenda Smith Myles, Melissa L. Trautman, and Ronda L. Schelvan

References

Bennett, M. (2020). *Life on the autism spectrum: A guide for the newly diagnosed*. Jessica Kingsley Publishers.

Bissonnette, B. (2010). *The Asperger's syndrome workplace survival guide: A neurotypical's secrets for success*. Future Horizons.

Finch, D. (2012). *The journal of best practices: A memoir of marriage, Asperger syndrome, and one man's quest to be a better husband*. Scribner.

Myles, B. S., Trautman, M. L., & Schelvan, R. L. (2013). *The hidden curriculum: Practical solutions for understanding unstated rules in social situations* (2nd ed.). AAPC Publishing.

Price, D. (2022). *Unmasking autism: Discovering the new faces of neurodiversity*. Harmony.

Robison, J. E. (2007). *Look me in the eye: My life with Asperger's*. Crown.

Silberman, S. (2015). *NeuroTribes: The legacy of autism and the future of neurodiversity*. Avery.

Soraya, L. (2013). *Living independently on the autism spectrum: What you need to know to move into a place of your own, succeed at work, start a relationship, stay safe, and enjoy life as an adult on the autism spectrum*. Adams Media.

Websites & Online Resources

1. Autistic Self Advocacy Network (ASAN) https://autisticadvocacy.org—Run by autistic individuals advocating for policy and social change.
2. NeuroClastic https://neuroclastic.com—Autism research, personal stories, and advocacy by autistic voices.
3. Wrong Planet https://wrongplanet.net—An online community with forums for autistic adults.
4. Autism Speaks Adult Services https://www.autismspeaks.org/adult-services—Resources on employment, housing, and daily living.
5. Neurodivergent Insights https://www.neurodivergentinsights.com—Infographics and educational tools on autism.
6. Thinking Person's Guide to Autism https://thinkingautismguide.com—Evidence-based information and real-life experiences.

Employment & Career Support

1. Neurodiversity Hub https://www.neurodiversityhub.org—Job resources and employer partnerships.
2. Specialisterne https://www.specialisterne.com—Helps autistic individuals find meaningful employment.
3. Microsoft Neurodiversity Hiring Program https://aka.ms/Neurodiversity
4. Google Autism Career Program https://disabilityin.org/autism-career-program—Google runs this program in partnership with Stanford Neurodiversity Project and Disability
5. Lime Connect https://www.limeconnect.com—Connects disabled professionals with job opportunities.

Independent Living & Life Skills

1. Autistic Women & Nonbinary Network (AWN) https://awnnetwork.org—Support for marginalized gender identities.
2. Autism Housing Network https://www.autismhousingnetwork.org—Resources for finding autism-friendly housing.
3. The ARC https://thearc.org—Advocacy for independent living and self-determination.
4. Asperger/Autism Network (AANE) https://www.aane.org—Coaching and support groups for autistic adults.

Mental Health & Therapy

1. TherapyDen https://www.therapyden.com—Directory of neurodivergent-friendly therapists.
2. Open Path Psychotherapy Collective https://www.openpathcollective.org—Affordable therapy options.
3. Psychology Today's Find an Autism Therapist https://www.psychologytoday.com/us/therapists/autism—Search for autism-informed therapists.
4. Autism Level UP! https://autismlevelup.com—Practical tools for emotional regulation and communication.

Online Courses & Skill Development

1. Skillshare https://www.skillshare.com—Various life and career skills courses.
2. Coursera https://www.coursera.org—Free and paid courses, including social skills training.
3. Udemy https://www.udemy.com—Self-paced learning for professional and personal growth.

Index

A

Acceptance xiii, xv, 25, 29, 99, 103, 109, 115, 125, 138, 206
Accommodations x–xii, xiv, 6, 12, 16, 20, 22, 25, 26, 30, 50–52, 66, 75–77, 81, 88, 99, 100, 104, 105, 163, 190, 192, 193, 196
Accountability 38, 148, 160, 169
Achievement 17, 24, 25, 29, 33, 50, 111, 123, 145, 151, 152, 154, 155, 157, 158, 182, 183, 195
Active listening 73, 83, 84, 89, 91, 92, 119
Adaptability ix, xvi, xviii, 1, 14, 22, 24, 41, 43, 45, 46, 52, 55–57, 89, 90, 107, 125, 127, 129, 136, 139–141, 187, 192
Adaptation xix, 35, 81, 83, 116, 118, 124, 125, 136, 181, 183, 199
Adaptive behavior 15
Adulthood xiii, 8, 11, 30, 104, 187, 199, 203
Advocacy xv, 13, 56, 107, 120, 121, 193
Alexithymia 98
Analytical skills 146, 148
Anchor routines 184
Anxiety x–xii, xiv, 6, 9, 11, 12, 22, 27, 32, 34–36, 41, 43, 45–50, 52, 55, 60, 63, 65, 74, 82, 89, 90, 98–102, 118–120, 124, 132, 134, 136, 139, 142, 164–168, 175, 177, 179, 182, 183, 191, 192, 198, 205
Aspirations 1, 96, 155, 159, 194, 195
Attention Deficit Disorder (ADHD) 162
Attention to detail attribution theory 136
Authenticity xiv, xx, 28, 31, 33, 54, 107–109, 114, 119, 125, 187, 188, 195, 207
Autism Spectrum Disorder (ASD) ix–xii, xiv–xx, xxiii, xxiv, 1–17, 19–24, 27–29, 32–36, 41–43, 45–55, 59–69, 73–91,

95–109, 113–127, 129–141, 145–148, 151, 153, 155, 156, 159, 160, 162–166, 168, 170, 173–184, 187–196, 199, 205, 206
Autistic identity ix, xiv

B

Behavior x, xi, xvii, 3, 6, 9–12, 15, 17, 19–24, 36, 38, 44, 61, 63, 65, 69, 75, 85, 86, 88, 97, 98, 100–103, 109, 114, 123, 124, 133, 146, 150–152, 160, 162, 165, 168, 175, 178, 206
Belonging ix, 7, 38, 51, 53, 113, 121, 126, 127, 183, 194, 196, 206
Black-and-white thinking 20, 136
Body language 3, 5, 7, 12, 61, 66, 69, 70, 73–76, 81, 86, 89, 97, 99, 121, 156, 157, 177, 185, 203
Boundaries ix, xv, 13, 14, 26, 28, 51, 52, 85, 96, 104, 105, 109, 111, 113, 118, 126, 127, 191–193, 195, 197
Burnout x, 12, 13, 26, 48, 97, 102, 103, 105, 109, 163, 164

C

Camouflaging xxiii, 12
Childhood 11, 99, 200
Clear communication 47, 75, 82, 88, 101, 106, 121, 123, 125, 190, 191
Clear expression 73, 84, 89
Cognitive behavioral therapy 7, 19, 76, 163
Cognitive flexibility 20
Cognitive load 49
Cognitive processing differences 20
Cognitive reframing 140

Cognitive strengths 145, 156
Cognitive tools 184
Collaboration xv, xvii, xviii, 23, 44, 80, 84, 88, 100, 133, 146, 150, 177
Commitment 29, 35, 42, 43, 45, 48, 50, 54, 55, 125, 128, 152, 161, 164, 168, 183
Communication ix, xii–xiv, xvi, 1, 3, 5, 8, 12, 15, 17, 22–24, 28, 29, 36, 44, 45, 49–51, 62, 66, 67, 69, 70, 73–92, 97, 100, 103, 105, 106, 108, 109, 113, 114, 116–119, 122–128, 130, 133, 134, 149, 150, 154, 156, 160, 190–193, 196
Communication breakdowns 90
Communication strategies 45
Community xi, xv, 5, 17, 25, 28, 33, 34, 36, 38, 51, 53, 57, 75, 76, 104, 106–110, 113, 120, 127, 130, 176, 181, 182, 184, 190, 191, 195, 206
Community support 108
Confidence xv, xviii, xix, 6, 8–10, 14–17, 20–22, 24, 25, 27, 29–38, 42–45, 53–56, 58, 68, 71, 73, 85, 90, 92, 96, 99, 102, 104–109, 119, 127, 130, 133–135, 137, 139–142, 145, 146, 151–159, 164, 166, 167, 169, 173–175, 181, 182, 184, 186, 187, 189, 193–196, 198
Conflict resolution 76, 84, 86–88
Consistency 27, 102, 160, 169–171, 180
Constructive feedback 34, 60, 82, 130, 176
Contextual challenges 162
Cooperative behavior 117, 133
Coping strategies xvii, 6, 7, 12, 13, 15, 47, 50, 53, 55, 153, 155, 192

Creativity xiv, 32, 37, 55, 108–110, 140, 176, 188, 193, 196

D

Daily affirmations 37
Daily practices 173, 179, 182
Decision-making 100, 122, 160, 163
Deep breathing 9, 14, 16, 31, 53, 55, 56, 90, 170
Diagnosis x, xv, xxiv, 8, 164, 187, 199, 206
Digital communication 67, 68
Digital environment 190, 196
Direct communication 28, 76, 104, 105, 118, 121, 126, 150

E

Emotional awareness 65, 68, 70, 183
Emotional connection 47, 48, 54, 97, 115, 122
Emotional dysregulation 164
Emotional expression xiii, 64, 65, 77, 97, 98, 115, 123, 147
Emotional labeling 65, 66, 68
Emotional reciprocity 76, 77, 97, 114, 115, 118, 126, 131, 147
Emotional regulation xvi, xviii, 2, 8, 9, 14, 15, 22, 31, 53, 55, 56, 63, 69, 71, 90, 161, 163, 171, 174, 185
Emotional resilience xvi, 7, 14, 53, 91, 162, 183, 185
Empathy xiv, xv, 1, 3, 4, 51, 53, 57, 59–65, 67–71, 74, 75, 77, 79, 86, 91, 97, 118, 128, 131, 147, 175, 185
Empowerment xiv, xv, xviii, xx, 19, 30, 31, 34, 36, 41, 44, 58, 125, 165
Environmental awareness 116
Environmental control 15
Executive dysfunction 99, 164

Executive function 8, 10, 48, 59, 61, 68, 77, 78, 98, 132, 160–164, 170, 176
Eye contact 4, 5, 12, 19, 28, 36, 62, 79, 80, 83, 84, 86, 99, 102, 106, 116, 119, 122, 124, 133, 137, 148, 149, 156

F

Facial expressions 7, 12, 60, 61, 66, 68–71, 73–77, 84, 86, 89, 99, 130–132, 149, 162, 176, 177
Family relationships 7
Fear of failure 135, 138–140, 143
Fear of success 152, 153
Feedback 23, 24, 34, 35, 38, 43, 47, 60, 74, 78, 82, 83, 86, 99, 100, 130, 137, 139, 142, 146, 149, 176, 177, 184
Flexibility 41, 48, 90, 92, 100, 122, 127, 164–166, 171, 175, 180, 183, 186, 192
Friendship fulfillment 197

G

Gaze interpretation 148
Goal setting xviii
Growth mindset 129, 136, 157, 195

H

Habits xi, 16, 37, 55, 70, 78, 90, 110, 160, 163–166, 168, 169, 171, 173, 179, 180, 183, 194
Harmful scripts xix, 10, 11, 19–21, 23, 26–35, 37, 39

I

Identity ix, xiv, 13, 33, 56, 95, 96, 104, 108, 135, 173, 187, 205
Imperfection 135, 138, 140

Inclusion xiii
Independence x, xiii, xviii, 2, 8, 10, 15, 90, 106, 110, 132, 159, 165, 169, 184, 188, 189, 196, 202
Individuality 54, 96, 102–104, 107–109, 113, 119, 188
Internalized beliefs 19
Interpersonal relationships 59, 70, 76, 150, 183
Islands of competence xvi, 32, 145, 153, 154, 156, 157

Job accommodations x
Job interviews x, 86, 92, 203
Job retention x
Joint attention 60, 68, 69, 80, 81, 100, 117, 133
Joint attention of behavior 5, 15, 80, 122, 124, 133, 149, 177
Joint attention of emotion 3, 4, 15, 47, 60, 61, 69, 76, 115, 121, 123, 131, 147

Learning accommodations 16
Learning from mistakes 136, 138, 167, 170
Literal interpretation 74
Logical reasoning 97, 147

Masking xiv, xxiv, 12, 13, 102, 103, 109, 163, 201–203, 205, 206
Mental energy 97, 163, 165, 170
Mental flexibility 61
Mentalizing 185
Mental state attribution 118
Mental states 3, 6, 60, 75, 76, 96, 97, 114, 115, 121, 123, 128, 130, 131, 146, 147, 174, 175

Mentorship 117, 140
Middle age 205
Mindfulness xviii, 7, 11, 13–16, 26, 31, 45, 53, 55, 56, 90, 91, 138, 162, 167, 174, 194
Mindset shift 12, 135, 136, 154
Miscommunication xxiii, 47, 101, 123, 148
Mistake avoidance 136
Mistakes 26, 65, 66, 81, 101, 106, 119, 129, 130, 134–143, 151, 160, 167–169, 195
Motivation xxiii, 17, 22, 42, 50, 54, 56, 58, 59, 96, 114, 137, 140, 151, 152, 154, 155, 157, 158, 160, 161, 166, 168, 169, 172

Negative narratives 11, 23, 27, 31, 32
Negative scripts xxiv, 10, 15, 19, 20, 22–24, 36
Neurodivergence 123, 191
Neurodiversity ix, xiii, xiv, xviii, 68, 95, 100, 103, 104, 106, 108–110, 121, 127, 163
Nonverbal communication 12, 66–68, 70, 79, 86, 87, 91, 99, 116, 148, 177

Organization xxiii, 8, 49, 77, 98, 107, 132, 190, 193, 198
Organizing 6, 10, 48, 98, 122, 134, 154, 165, 170, 171, 189, 194
Overload 31, 163, 164, 171

Parent-child relationships 123–125
Parenting x, xiii, xviii, 123–127, 204
Peer support 30

Perfectionism 134, 140, 141, 165, 168
Personal boundaries ix
Personal control 7, 9, 11, 12, 15, 17, 42, 44, 52, 56
Personal growth xviii, 9, 19, 31, 37, 51, 57, 106, 120, 129, 145, 159, 165, 170
Personal narrative 108
Personal values 42, 55, 95, 104, 110, 197
Perspective-taking 76, 131
Planning 4, 6, 48, 49, 57, 61, 77, 98, 122, 123, 127, 128, 132, 147, 158, 168, 170, 174, 176, 185, 192, 194
Predictability 7, 23, 31, 125, 127, 132, 164, 165, 170, 174, 179, 184, 191
Prioritization 8, 48, 49, 147, 148, 158
Problem-solving xviii, 6, 7, 12, 42, 44, 55, 77, 98, 104, 117, 139, 140, 142, 147, 154, 161
Professional development 9

R

Reflection 11, 28–30, 35, 36, 38, 57, 58, 70, 71, 97, 105, 128, 135–137, 139, 141, 170, 173, 175, 179, 184, 193, 199
Reframing 11, 15, 20, 23, 27, 29–31, 35, 51, 52, 54–56, 104, 106, 110, 129, 136, 137, 139, 141, 142, 158
Reframing negative narratives 109
Regulation strategies 22, 31, 68, 69, 90
Relationship building xi
Resilience ix, xii–xx, xxiii, xxiv, 1, 2, 6–11, 13–20, 24, 25, 28–32, 35, 36, 39, 41–45, 50–58, 73, 88, 90, 95, 96, 106–108, 113, 119, 120, 125, 127, 130, 135, 136, 139–141, 145, 151, 153–155, 157, 158, 160, 170, 173–187, 189, 192, 194, 195, 197, 199, 206, 207
Resilient mindset xvi, xviii, xix, 1, 2, 9–11, 13, 174
Reward system 158
Rigid thinking x, 20, 134, 139, 141
Role-playing xix, 23, 65, 66, 68, 82, 90, 92, 130, 133, 142
Romantic relationships x, xiii, 28, 76, 97, 121–123, 127
Routine xxiv, 4, 7, 8, 17, 23, 31, 42–45, 48, 52–56, 91, 99, 123, 125, 127, 131, 132, 155, 156, 160–173, 176, 178–184, 186, 189–192, 194, 197, 200, 205
Routine changes 122

S

Sarcasm 6, 66, 74, 81, 101, 117, 131, 134
Sarcasm detection 66
Self-acceptance xv, 25, 26, 57, 95–97, 99–111, 121, 126, 129, 135, 140, 141, 155, 199, 207
Self-advocacy xiv, 8, 12, 15, 17, 50, 52, 55, 90, 105–107, 155, 192, 193
Self-awareness ix, xiv, xvi, 2, 7–10, 15, 17, 27, 32, 53, 55, 66, 71, 88, 109, 128, 152, 162, 174, 183, 192
Self-care 45, 46, 171, 179, 183, 197
Self-compassion 16, 25, 26, 30, 36, 99, 103, 129, 138, 142, 143, 181, 188
Self-confidence xviii, 10, 39, 44, 88, 91, 102, 129, 145, 154
Self-control 159–161, 170

Self-criticism 25, 26, 57, 98, 106, 129, 138, 140, 141
Self-determination 211
Self-discipline xvii, xviii, 159–161, 163–165, 168, 169, 172
Self-discovery xviii, 154, 187, 195
Self-doubt 6, 22, 25, 96, 99, 102, 106, 119, 151, 154
Self-identity ix, 109, 111
Self-regulation 8, 13, 38, 53, 63, 69, 160
Sensory needs 107, 108, 189
Sensory overload x, xxiv, 2, 7, 9–11, 13, 14, 19, 21, 24, 25, 30, 53, 56, 62, 90, 155, 163, 189, 190, 195, 196, 205
Sensory planning xxiii
Sensory processing xi, 59, 62, 68, 69
Sensory regulation 17, 164
Sensory sensitivities x, xii, xxiii, 1, 8, 21, 30, 42–45, 52, 62, 69, 74, 79, 83, 98, 100, 103, 120, 164, 182
Shared activities 114, 122, 124, 191
Social ambiguity 150
Social anxiety xi, 9, 33, 47, 49, 74, 97, 154
Social awareness 70, 77
Social challenges 7, 9, 21, 33, 101, 103
Social-cognitive differences 161
Social communication x, xii, 44, 100, 177, 182
Social complexity 68, 150, 184
Social connection xiii, xvi, 66, 74, 91, 96, 115, 120, 190, 191, 197
Social cues x, xiii, xxiii, 3, 14, 15, 19, 21, 27, 41, 64, 67, 74, 83, 85, 87, 89, 92, 97, 99, 106, 114, 123, 126, 130, 133, 134, 146, 156, 176, 205
Social dynamics ix, 5, 51, 79, 81, 101, 134, 178, 184

Social expectations 3, 7, 8, 16, 47, 50, 77, 97, 114, 124, 195, 202
Social fatigue 102, 175
Social flowcharts 178, 184
Social fulfillment 191, 197
Social information processing 118
Social interaction xi, xvi, 1, 2, 5–11, 13, 14, 20, 21, 23, 24, 27–29, 31, 32, 38, 41, 42, 44, 45, 47, 49, 60–66, 68–71, 73, 75, 76, 78, 80–82, 86, 88–90, 92, 99–102, 106, 114, 116, 119, 125, 127, 128, 130–133, 135, 139, 141, 142, 146–150, 155, 156, 184, 185, 199, 205
Social learning 101, 102
Social navigation xii, 1
Social needs 113, 126, 191
Social reciprocity 115
Social rules 5, 80, 84, 89, 202
Social scripts 27, 47, 71, 115, 127, 134, 175
Social skills 9, 11, 14, 76, 77, 101, 119, 130, 157
Social stories 64, 65, 68, 89, 131, 186
Social strategies 146, 147, 157
Social support 16, 127
Social support networks 120
Solitude 100, 201, 202
Stigma 119
Strength-based approach xiv
Strengths ix, x, xiii–xvii, xix–xxi, xxiii, 1, 2, 4, 12, 16, 20, 24, 25, 27–33, 36, 37, 41, 43–45, 50, 54, 55, 57, 58, 92, 95, 96, 98, 99, 103–111, 114, 117, 119, 121, 124–127, 134, 145–150, 153–158, 172–174, 180, 182, 184, 186–188, 190, 193–197, 201
Stress x, xii, xiv, 5–7, 9, 12–16, 31, 32, 41, 42, 45–56, 58, 65, 71, 90, 95, 98–100, 114, 120,

122, 129, 132, 141, 143, 148, 150, 153, 163–168, 170, 171, 173, 185, 189, 190, 192, 194, 196–198
Stress hardiness 42–46, 50, 54
Stress management 44, 58
Structure x, 20, 43, 45, 49, 82, 123, 124, 127, 165, 166, 170, 176–179, 186, 191, 192, 200, 202, 205
Structured routines 15, 46, 48, 104, 122, 132, 164, 183, 188
Structured support 77, 82, 99, 130, 132, 141, 175
Structured thinking 150
Success ix, xii, xiii, xv, xvii, xix, 2, 3, 6, 8–10, 15, 22–25, 28, 29, 31–33, 35, 37, 38, 44, 51, 88, 89, 91, 105, 107, 121, 128–130, 132, 136, 139, 140, 143, 145–148, 151–158, 168, 172, 174, 179, 182, 186, 188, 190, 193, 195
Supportive coaching 9
Supportive environments xviii, 47, 86, 118, 129, 130, 140, 189, 190, 196
Support networks xiii, 53, 55, 57, 120, 127, 180, 181, 184, 192, 194
Support systems x, xi, xiii, xvi, 2, 6, 9, 13, 15, 91, 113, 169, 174, 184

Task initiation 58
Task management 158, 163, 166
Teamwork 80, 88, 100, 117, 118, 133, 149, 150, 177

Theory of Mind (ToM) 3, 15, 46, 51, 52, 59, 60, 68, 69, 71, 75, 76, 97, 114, 130, 131, 175
Time management 8, 166, 168, 169
Tone of voice 7, 59, 61, 66, 68, 69, 73, 75, 76, 81, 85, 86, 131, 177
Transformative change 35
Transitions xiii, 48, 82, 105, 124, 125, 137, 148, 153, 161, 164, 170, 182, 191, 192, 197
Trust 84, 88, 92, 114, 118, 123, 127, 140, 147, 151, 152, 169, 180, 195, 198

Undiagnosed autism 199

Values xiii, xviii, xxiv, 24, 28, 29, 31, 33, 37, 43, 53, 57, 68, 95, 103–108, 110, 128, 141, 148, 155, 159, 169, 172, 179, 188, 189, 195–197
Verbal affirmation 80, 121, 122, 124, 126, 147
Viewpoint diversity 1
Visual schedules 49, 118, 122, 124, 127, 132, 161, 170, 192
Visual supports 89
Vulnerability 20

Workplace challenges xi, 42
Workplace integration 134

GPSR Compliance
The European Union's (EU) General Product Safety Regulation (GPSR) is a set of rules that requires consumer products to be safe and our obligations to ensure this.

If you have any concerns about our products, you can contact us on

ProductSafety@springernature.com

In case Publisher is established outside the EU, the EU authorized representative is:

Springer Nature Customer Service Center GmbH
Europaplatz 3
69115 Heidelberg, Germany

www.ingramcontent.com/pod-product-compliance
Lightning Source LLC
LaVergne TN
LVHW020132080526
838202LV00047B/3922